Lecture Notes in Mathematics

Edited by A. Dold and B. E

T0240101

625

J. P. Alexander
P. E. Conner
G. C. Hamrick

Odd Order Group Actions and Witt Classification of Innerproducts

Springer-Verlag
Berlin Heidelberg New York 1977

Authors
John P. Alexander
Gary C. Hamrick
Department of Mathematics
University of Texas
Austin, TX 78712/USA

Pierre E. Conner
Mathematics Department
LSU
Baton Rouge, LA 70803/USA

AMS Subject Classifications (1970): 57E10, 57D85, 10C05

ISBN 3-540-08528-9 Springer-Verlag Berlin Heidelberg New York
ISBN 0-387-08528-9 Springer-Verlag New York Heidelberg Berlin

This work is subject to copyright. All rights are reserved, whether the whole or part of the material is concerned, specifically those of translation, reprinting, re-use of illustrations, broadcasting, reproduction by photocopying machine or similar means, and storage in data banks.

Under § 54 of the German Copyright Law where copies are made for other than private use, a fee is payable to the publisher, the amount of the fee to be determined by agreement with the publisher.

© by Springer-Verlag Berlin Heidelberg 1977

Printed in Germany

Printing and binding: Beltz Offsetdruck, Hemsbach/Bergstr.
2140/3140-543210

Introduction

The present set of notes is intended as an exposition of the research program with which the authors have been principally concerned over the past four years. The phenomenon which has been the object of our interest arises from the observation that associated to a right action of a finite group as a group of orientation preserving diffeomorphisms on a closed oriented even dimensional manifold (M^{2m},π) there is induced an integral representation $(\pi, H^m(M;Z)/TOR)$ on the quotient of the middle dimensional integral cohomology group modulo the torsion subgroup. Furthermore π acts as a group of isometries with respect to the $(-1)^m$-symmetric integral valued unimodular bilinear form induced on $H^m(M;Z)/TOR$ by cup-product. Thus first we ask algebraically for a natural classification of integral representations of π as a group of isometries with respect to a (skew) symmetric Z-valued innerproduct and then we ask topologically for relations between the algebraic classification of $(\pi, H^m(M;Z)/TOR)$ and the fixed point data associated with the various elements of π.

The appropriate algebraic definition arises directly form the 1937 paper of Witt [W]. A study of the object appropriate to our purposes was initiated by Andreas Dress [D1,2,3].

Suppose (π,V,b) is a triple consisting of

1) a finitely generated free abelian left $Z(\pi)$-module (π,V)

2) a (skew) symmetric integral valued inner product structure (V,b) with respect to which π acts as a group of isometries in the sense $b(gv,gw) \equiv b(v,w)$.

Such a triple (π,V,b) is Witt equivalent to 0 if and only if there is a π-invariant submodule $N \subset V$ for which

$$N = N^{\perp} = \{v \mid v \in V, b(v,N) = 0\} .$$

Further, (π, V, b) is Witt equivalent to (π, V_1, b_1) if and only if the orthogonal sum $(\pi, V \oplus V_1, b \oplus (-b_1))$ is Witt equivalent to 0. This will be shown to be an equivalence relation. The groups $W_0(Z,\pi)$ and $W_2(Z,\pi)$ are made up respectively of equivalence (Witt) classes of symmetric and skew-symmetric triples (π, V, b). Addition is by orthogonal sums and $(\pi, V, -b)$ yields additive inverses. Finally

$$W_*(Z,\pi) = W_0(Z,\pi) \oplus W_2(Z,\pi)$$

is made into a $(Z/4Z)$-graded commutative ring with identity. We caution the reader that Dress uses $GW(Z,\pi)$ to denote $W_0(Z,\pi)$.

Use of the Witt classification allows us to define a homomorphism from the oriented bordism ring of all right π-actions

$$\omega: \mathcal{O}^{SO}_{2_*}(\pi) \rightarrow W_*(Z;\pi)$$

which preserves degree mod 4. This assigns to (M^{2m}, π) the Witt class of the induced $(\pi, H^m(M;Z)/TOR, b)$. This is a multiplicative homomorphism and we refer to it as the Witt invariant of $[M^{2m}, \pi]$.

In fact Z can be replaced by any Dedekind domain or by a field. If the real number field is employed then the resulting homomorphism of $\mathcal{O}^{SO}_{2_*}(\pi)$ into $W_*(\mathbb{R},\pi)$ is determined by the Atiyah-Singer-Segal G-Signature Theorem [A-S]. Our problem then splits into two parts:

1. Is it possible to effectively compute $W_*(Z,\pi)$ algebraically?
2. Is there additional topological information available from the use of integral rather than real coefficients?

The Dress induction techniques reduce, for π a finite odd order group, the determination of $W_*(Z,\pi)$ to the calculation for $\pi = C_n$, a cyclic group of odd order. The algebra in the present set of notes is primarily devoted to a determination of $W_*(Z,C_n)$, at least additively.

We find that if π is a p-group, p an odd prime, then $W_*(Z,\pi)$ contains no torsion and hence $W_*(Z,\pi) \to W_*(\mathbb{R},\pi)$ is a monomorphism. By contrast, however, if C_n is a cyclic group of odd composite order then $W_*(Z,C_n)$ may contain torsion elements. In fact $W_*(Z,C_{15})$ contains a single non-trivial element of order 2. The kernel of $W_*(Z,\pi) \to W_*(\mathbb{R},\pi)$ is the torsion ideal, which Dress shows is also the ideal of nilpotent elements.

Thus topologically, for π of odd order, we must, and do, find invariants in terms of the fixed point data of (M^{2m},π) which, when coupled with the Atiyah-Singer-Segal G-Signature Theorem, will determine $w([M^{2m},\pi])$ uniquely in $W_*(Z,\pi)$.

Let us now go into the plan of the chapters. In chapter I we present the basic definitions and results. This will be familiar to a number of readers, of course. The single most important concept we draw to our reader's attention is that of an S-anisotropic representative of a Witt class. This is the generalization of the classical notion of an anisotropic quadratic form over a field. Indeed later we shall generally refer to S-anisotropic simply as anisotropic.

Chapter II is our first venture into topology. For $\pi = C_{p^n}$, a cyclic group of odd prime power order, a finite symmetric inner product can be defined on the cohomology group

$$H^m(C_{p^n};H^m(M;Z)/TOR)$$

producing an invariant in the ordinary Witt ring $W(F_p)$ over the finite field of integers mod p. This is effectively computed in terms of the signatures of fixed point sets of the subgroups of C_{p^n} and several corollaries are then drawn from this.

Chapter III leads up to the determination of the order of the torsion subgroup of $W_0(Z,C_n)$, n odd. (It turns out that this torsion subgroup is a finite elementary abelian 2-group which can be non-trivial only if n is composite.) It is in this chapter that we begin to call upon some algebraic number theory. If $Q(\lambda)$ is the n-cyclotomic extension of Q then with respect to the involution complex conjugation there are the Witt groups $\mathcal{H}_0(Q(\lambda))$, $\mathcal{H}_2(Q(\lambda))$ of respectively Hermitian and skew-Hermitian innerproduct spaces over $Q(\lambda)$. We study the groups $\mathcal{H}_*(Q(\lambda))$ and their relation to the following. If $N \geq 1 \in Z$ then $Z(\lambda/N) \subset Q(\lambda)$ is the integral closure of $Z(1/N)$ in $Q(\lambda)$. Then $Z(\lambda/N)$ is a Dedekind domain stable under complex conjugation so there is $\mathcal{H}_*(Z(\lambda/N))$, which we proceed to compute. Note for $N = 1$ we have $Z(\lambda)$ the ring of algebraic integers in $Q(\lambda)$ and among other things we show $\mathcal{H}_*(Z(\lambda))$ is without torsion. By contrast if $N > 1$ then in general $\mathcal{H}_*(Z(\lambda/N))$ will have torsion.

Now we come to the critical chapter IV. This is devoted to the algebra needed to detect elements in $W_*(Z, C_n)$, n odd. The multisignature will detect everything but the torsion subgroup. First we prepare ourselves by comments about the Grothendieck representation ring of π. Specifically for $GR(F_p, C_n)$ we are concerned with the involution of this ring defined by dualizing a representation of C_n over the finite field F_p. This yields a C_2-module structure on $GR(F_p, C_n)$ and hence $H^2(C_2; GR(F_p, C_n))$ is defined.

Now for each prime $p|n$ and $1 \leq s \leq$ multiplicity of p in n there is introduced a homomorphism

$$\alpha(p^s, -): \; W_*(Z, C_n) \to H^2(C_2; GR(F_p, C_n)) \; .$$

In essence for a (C_n, V, b) there is a finite form on $H^1(C_{p^s}; V)$ with respect to which C_n still acts as a group of isometries (with $C_{p^s} \subset C_n$ acting trivially) and this representation is self-dual so there results an element $\alpha(p^s, \langle C_n, V, b\rangle) \in H^2(C_2; GR(F_p, C_n))$. All the α-invariants together with the multisignature will detect $W_*(Z, C_n)$, C_n a cyclic group of odd order.

In Chapter V we reach the Torsion G-Signature Theorem. Thus for an action (M^{2m}, C_n) there is a family of torsion signatures which are themselves determined by fixed point data and which in turn determine the α-invariants. It should be noted that we have not been able to make a direct connection between α-invariants and the topology. For this reason the torsion signatures (actually mod 2 invariants) live in a quotient of $H^2(C_2; GR(F_p, C_n))$. It is true that the vanishing of all torsion signatures does imply the vanishing of α-invariants.

Chapter VI is really designed to exposite a bit about the relation of $W_*(Z,\pi)$ to the Wall surgery obstruction groups $L^h(\pi)$ and $L^P(\pi)$. We content ourselves with $\pi = C_{p^n}$ a cyclic group of odd prime power order. Actually our notion of Witt equivalent to zero does not, over Z, coincide with the classical idea of a hyperbolic quadratic form. Suppose (π,N) is an integral representation of π on a finitely generated free abelian group. Then the hyperbolic form associated to (π,N) is

$$(\pi,N) \oplus (\pi,\mathrm{Hom}_Z(N,Z))$$

with innerproduct

$$[(x,\varphi),(y,\psi)] = \psi(x) + \varphi(y) .$$

Thus we may ask if metabolic implies hyperbolic? The integral form must be even (of type II) to have a chance at hyperbolic. We show in fact that for a cyclic group of odd prime order any (C_p,V,b) which is Witt equivalent to zero and for which b is even must in fact be of hyperbolic type.

This does not generalize even to cyclic groups of odd prime power order. However if (C_{p^n},V,b) is assumed to be even, Witt equivalent to 0, and V projective as a $Z[C_{p^n}]$ module then it is shown (C_{p^n},V,b) is hyperbolic. We go on to obtain a special case of a theorem of Bak [B] which relates $L^h(\pi)$ to $L^P(\pi)$.

Background material is generally quoted to an extent sufficient to our purposes. For the reader who wishes a more detailed account we shall mention some sources from which we have drawn. The results from oriented bordism employed in chapter II are drawn from [C-F,MOP]. A complete account of Hilbert symbols can be read in [O'M., secs. 63B-71]. We simply state the theorems needed in chapter III. A complete account, for our purposes, of cyclotomic number fields is given in the last chapter of [We]. The 1960 Annals paper of Swan [Sw 1] introduces his fundamental results on the Grothendieck representation ring of a finite group.

On the algebraic side we have been most heavily influenced by the Milnor-Husemuller presentation of modern Witt theory [M-H]. We shall of course refer frequently to this book and we adopt in particular their convention in the use of the term innerproduct space.

We have aimed for as complete an exposition as possible, both algebraic and topological, of the Witt theory $W_*(Z,\pi)$ for finite groups of odd order. By contrast there is an open field for investigation in connection with even order groups. We could refer to [A] and [G] for actions of the cyclic group of order 2, C_2. A great deal remains to be done even for actions of the cyclic groups C_{2^k}.

A most extensive discussion of Hermitian forms in general will be found in the Wall references. We wish to acknowledge the interest shown by Andreas Dress in this work. Additional developments and applications of the Witt classification technique will be found in Stoltzfus' paper describing the integral knot concordance group [St]. We should also express our thanks to J. W. Vick who collaborated with us on earlier work in this area.

Chapter I
Algebraic Preliminaries

1. The Witt group $W_*(D,S)$

In this first section we shall introduce those algebraic definitions the study of which will be the subject of these notes. Let D be either a Dedekind domain or a field. Then S will denote an associative linear algebra over D with identity. If $1 \in S$ is this identity element then we assume that $D \to S$ given by $d \to d\cdot 1$ is an embedding of D into the center of S. Thus $D \subset S$. We shall also assume that S is equipped with an anti-involution $s \to \hat{s}$ that reduces to the identity on $D \subset S$.

(1.1) Definition: A representation of S, denoted by (S,V), is a left S-module which, with respect to the induced structure, is a finitely generated projective D-module. An orthogonal representation of S is a triple (S,V,b) wherein

1) (S,V) is a representation of S
2) (V,b) is a D-valued symmetric innerproduct space structure on V

3) *for* $s \in S$, v, w *in* V

$$b(sv,w) \equiv b(v,\hat{s}w) .$$

The notion of *skew-symmetric representation* is similarly defined by requiring (V,b) be a skew-symmetric innerproduct rather than a symmetric one. The basic example we have in mind for S is the integral group ring of a finite group. To say $(S,V,b) \simeq (S,V_1,b_1)$ means that there is an S-module isometry.

If (S,V,b) is an orthogonal (skew-symmetric) representation then for an S-submodule $W \subset V$ we define the complement by

$$W^{\perp} = \{v \mid v \in V, b(v,w) \equiv 0 \text{ all } w \in W\} .$$

Obviously $W^{\perp}$ is also an S-submodule. It is also true

(1.2) Lemma: *For an S-submodule $W \subset V$ the complement $W^{\perp}$ is a D-module summand of V.*

Proof: Since D is a Dedekind domain and V is finitely generated and D-projective it is enough to show the quotient $V/W^{\perp}$ is torsion free. If $v \in V$ for which there is a $d \neq 0 \in D$ with $dv \in W^{\perp}$ then for all $w \in W$ $b(dv,w) \equiv db(v,w) \equiv 0$. Plainly $b(v,w) \equiv 0$ and $v \in W^{\perp}$. □

Now we come to a key lemma which, in various guises, will appear frequently.

(1.3) Lemma: *If (S,V,b) is an orthogonal (skew-symmetric) representation and if $W \subset V$ is an S-submodule which is a D-module direct summand of V then $(W^{\perp})^{\perp} = W$.*

Proof: We may in fact ignore S entirely for this proof. Define $P: V \to \mathrm{Hom}_D(W,D)$ by

$$P(v)(w) \equiv b(v,w) .$$

This is an epimorphism because W is a D-summand. So we have a short exact sequence

$$0 \to W^{\perp} \xrightarrow{i} V \xrightarrow{P} \mathrm{Hom}_D(W,D) \to 0 .$$

All modules involved are finitely generated and D-projective and hence there is a dual short exact sequence

$$0 \to \mathrm{Hom}_D(\mathrm{Hom}_D(W,D)) \xrightarrow{P^*} \mathrm{Hom}_D(V,D) \xrightarrow{i^*} \mathrm{Hom}_D(W^{\perp},D) \to 0 .$$

Next let us remember that we also have

$$0 \to (W^{\perp})^{\perp} \xrightarrow{j} V \xrightarrow{\tilde{P}} \mathrm{Hom}_D(W^{\perp},D) \to 0$$

where $\tilde{P}(v)(u) = b(v,u)$, $u \in W^{\perp}$. We have $\mathrm{id}\colon \mathrm{Hom}_D(W^{\perp},D) \simeq \mathrm{Hom}_D(W^{\perp},D)$ and $\mathrm{Ad}\colon V \simeq \mathrm{Hom}_D(V,D)$. We need to define a homomorphism

$$\alpha\colon (W^{\perp})^{\perp} \to \mathrm{Hom}_D(\mathrm{Hom}_D(W,D),D) .$$

Suppose $w_1 \in (W^{\perp})^{\perp}$ and $\varphi \in \mathrm{Hom}_D(W,D)$. We put

$$\alpha_{w_1}(\varphi) = b(v,w_1) \in D$$

where $v \in V$ is any element with $P(v) = \varphi$. This is well defined for if $P(v') = \varphi$ also then $v - v' \in W^{\perp}$ and $b(v-v',w_1) = 0$ because $w_1 \in (W^{\perp})$. In this way we arrive at a commutative diagram

$$\begin{array}{ccccccc}
0 \to \mathrm{Hom}_D(\mathrm{Hom}_D(W,D),D) & \xrightarrow{P^*} & \mathrm{Hom}_D(V,D) & \xrightarrow{i^*} & \mathrm{Hom}_D(W^{\perp},D) \to 0 \\
\uparrow \alpha & & \simeq \uparrow \mathrm{Ad} & & \uparrow \mathrm{id} \\
0 \longrightarrow (W^{\perp})^{\perp} & \longrightarrow & V & \longrightarrow & \mathrm{Hom}_D(W^{\perp},D) \to 0
\end{array}$$

and hence

$$\alpha: (W^{\perp})^{\perp} \simeq \mathrm{Hom}_D(\mathrm{Hom}_D(W,D),D) .$$

Obviously $W \subset (W^{\perp})^{\perp}$ and if $w \in W$ then $\alpha_w(\varphi) = b(v,w) = P(v)(w) = \varphi(w)$. This means that the restriction of α to W is the canonical isomorphism of W with its double dual. This shows $W = (W^{\perp})^{\perp}$ and completes the proof of (1.3). □

Let us say a few more useful things about taking complements.

(1.4) Lemma: For any S-module $W \subset V$, $W \subset (W^{\perp})^{\perp}$ and the quotient is a finitely generated torsion D-module isomorphic to the torsion submodule of V/W.

Proof: Let

$$W_1 = \{v | v \in V, \quad d \neq 0 \in D \quad dv \in W\} .$$

Then W_1 is a D-module summand with $W^{\perp} = W_1^{\perp}$ so that $W \subset (W^{\perp})^{\perp} = W_1$ and W_1/W is the torsion submodule of V/W. □

(1.5) Lemma: For any two S-submodules W, W_1 in V

$$(W + W_1)^{\perp} = W^{\perp} \cap W_1^{\perp} .$$

Proof: Obvious from the definitions. □

(1.6) Lemma: If W, W_1 are also both D-module summands of V, then

$$W^{\perp} + W_1^{\perp} \subset (W \cap W_1)^{\perp}$$

with quotient a finitely generated torsion D-module.

Proof: Write

$$(W_1^\perp + W_2^\perp)^\perp = (W_1^\perp)^\perp \cap (W_2^\perp)^\perp = W_1 \cap W_2 .$$

Then $W_1^\perp + W_2^\perp \subset (W_1 \cap W_2)^\perp$ with quotient as described by (I.1.5). □

To have an equality we would need to know $W_1^\perp + W_2^\perp$ is a D-module summand of V, but this is not always the case.

Fundamental to Witt equivalence is the notion of metabolic.

(1.7) Definition: An orthogonal (skew-symmetric) representation (S,V,b) is metabolic if and only if there is an S-submodule $N \subset V$ for which $N = N^\perp$.

Such an S-submodule is called a metabolizer. Such a metabolizer is always a D-module summand.

(1.8) Lemma: Let (S,V,b) be a metabolic representation and let L be an S-submodule which is maximal with respect to $L \subset L^\perp$, then L is a metabolizer.

Proof: We claim L is a D-module summand, for if

$$L_1 = \{v \mid v \in V, \quad d \neq 0 \in D \quad dv \in L\}$$

then $L_1^\perp = L$ and obviously $L_1 \subset L_1^\perp$ so by maximality $L_1 = L$.

Next let $N \subset V$ be any metabolizer for (S,V,b). We claim $L^\perp \cap N \subset L$. Look at $L + L^\perp \cap N$ and verify this is contained in its own orthogonal complement so by maximality

$$L = L + L^\perp \cap N .$$

From this it follows that

$$L \subset L^\perp \subset (L^\perp \cap N)^\perp .$$

Apply (I.1.6) so that $L + N \subset (L^\perp \cap N)^\perp$ with quotient a torsion D-module. But then $L^\perp / L^\perp \cap (L+N)$ is also a finitely generated torsion D-module. It is immediately seen, however, that $L^\perp \cap (L+N) = L$. Thus $L = L^\perp$ since L is a D-summand.

(1.9) Corollary: *Let (S,V,b) be a metabolic representation. If $W \subset V$ is an S-submodule for which $W \subset W^\perp$ then W is contained in a metabolizer.*

Proof: Appealing to the A.C.C. for D (and hence S) submodules, W lies in an S-submodule which is maximal with respect to being contained in its complement. □

(1.10) Theorem: *Suppose (S,V,b) is a representation and (S,M,β) is a metabolic representation. If $(S,V,b) \oplus (S,M,\beta) = (S,V \oplus M, b \oplus \beta)$ is metabolic then (S,V,b) is also metabolic.*

Proof: Let $W \subset M$ be a metabolizer for (S,M,β). If $W \subset V \oplus M$ as $(0,W)$ then $W \subset W^\perp$ with respect to $b \oplus \beta$. Since $(S,V \oplus M, b \oplus \beta)$ is also metabolic, there is a metabolizer $N \subset V \oplus M$ for which $W \subset N$.

It is $L = V \cap N$ which is the metabolizer for (S,V,b). If $(v,x) \in N$ then for all $w \in W$, $(0,w) \in N$ and $b(v,0) + \beta(x,w) = \beta(x,w) = 0$. Thus $x \in W^\perp = W$. In particular $(v,0) \in N$ also. Thus $N = L \oplus W$ and so $L = L^\perp$ with respect to (V,b). □

Finally we are prepared to introduce Witt equivalence

(1.11) Definition: *If (S,V,b) and (S,V_1,b_1) are orthogonal (skew-symmetric) representations then*

$$(S,V,b) \sim (S,V_1,b_1)$$

if and only if

$$(S,V,b) \oplus (S,V_1,-b_1) = (S,V \oplus V_1, b \oplus (-b_1))$$

is metabolic.

We note that $(S,V \oplus V,b \oplus (-b))$ is metabolic with the diagonal as metabolizer. Thus $(S,V,b) \sim (S,V,b)$. Symmetry is left as an exercise. The crucial issue is transitivity. Thus suppose

$$(S,V,b) \sim (S,V_1,b_1)$$
$$(S,V_1,b_1) \sim (S,V_2,b_2)\ .$$

Form

$$(S,V,b) \oplus (S,V_2,-b_2) \oplus (S,V_1,-b_1) + (S,V_1,b_1)$$
$$\simeq (S,V \oplus V_1, b \oplus (-b_1)) \oplus (S,V_1 \oplus V_2, b_1 \oplus (-b_2))\ .$$

It follows from (I.1.10) that $(S,V \oplus V_2, b \oplus (-b_2))$ is metabolic.

Of course if $(S,V,b) \simeq (S,V_1,b_1)$ then $(S,V,b) \sim (S,V_1,b_1)$. The graph of the S-module isometry is a metabolizer.

Now by $\langle S,V,b\rangle$ we denote the Witt class of (S,V,b), by $W_0(D,S)$ the collection of all orthogonal Witt classes and by $W_2(D,S)$ the collection of skew-symmetric Witt classes. These are abelian groups with

$$\langle S,V,b\rangle + \langle S,V_1,b_1\rangle$$
$$= \langle S,V \oplus V_1, b \oplus b_1\rangle$$

and $$-\langle S,V,b\rangle = \langle S,V,-b\rangle\ .$$

Thus in these groups the 0 element is the class of all metabolics.

If $S = D$ we have $W_0(D,D) = W(D)$ the ordinary Witt group of D-module inner-product spaces. This is computable if D is the ring of integers in an algebraic number field ([M-H]). Also $W_2(D,D) = 0$, if $ch(D) \neq 2$.

For $S = D(\pi)$, the group ring of a finite group with $\hat{g} = g^{-1}$, we can regard $(D(\pi),V,b)$ as a π-module V for which π acts as a group of isometries with respect to b.

(1.12) __Lemma__: __Suppose__ W __is a__ S-__submodule of__ (S,V,b) __for which__ $W \subset W^{\perp}$. __If__ W __is a__ D-__module summand of__ V __then__ $(S,W^{\perp}/W,\beta)$ __is naturally defined and__

$\langle S,V,b\rangle$ *is Witt equivalent to* $\langle S,W^\perp/W,\beta\rangle$.

Proof: If $\nu\colon W^\perp \to W^\perp/W$ is the quotient homomorphism then let $\beta(\nu(v),\nu(u)) = b(v,u)$ for v,u in $W^\perp$. This is well defined. Now an element of $\mathrm{Hom}_D(W^\perp/W,D)$ can be thought of as a homomorphism $\varphi\colon W^\perp \to D$ which is 0 on W. Since $W^\perp$ is a summand of V we can find a $u \in V$ such that $b(v,u) = \varphi(v)$ for all $v \in W^\perp$. Since $\varphi|W \equiv 0$ it follows that $u \in W^\perp$. This shows $\mathrm{Ad}\colon W^\perp/W \to \mathrm{Hom}_D(W^\perp/W,D)$ is an epimorphism. If the adjoint had a kernel there would be a $u \in W^\perp$ with $b(v,u) = 0$, all $v \in W^\perp$. But then $u \in (W^\perp)^\perp = W$ and so $\nu(u) = 0 \in W^\perp/W$.

Now $\nu\colon W^\perp \to W^\perp/W$ is an S-module homomorphism. Its graph in $(S,V \oplus W^\perp/W,b \oplus (-\beta))$ is a metabolizer. Denoting the graph by N it is clear that $N \subset N^\perp$. We need to see $N^\perp \subset N$. Thus suppose $(x,\nu(y)) \in N^\perp$ where $x \in V$ and $y \in W^\perp$. Then for all $v \in W^\perp$

$$b(x,v) - \beta(\nu(y),\nu(v)) = 0 .$$

Since $\beta(\nu(y),\nu(v)) = b(y,v)$ this means $b(x-y,v) = 0$ for all $v \in W^\perp$. Thus $x - y \in (W^\perp)^\perp = W$ and so $x \in W^\perp$ since $y \in W^\perp$ and $\nu(x) = \nu(y)$.

This shows $\langle S,W^\perp/W,\beta\rangle = \langle S,V,b\rangle$. □

(1.13) ***Definition***: *An orthogonal (skew-symmetric) representation* (S,V,b) *is S-anisotropic if and only if for every S-submodule* $W \subset V$

$$W \cap W^\perp = \{0\} .$$

This is the generalization of what is classically meant by the term anisotropic form.

(1.14) ***Theorem***: *Every Witt class admits an S-anisotropic representative. Furthermore, if D is a field this representative is unique up to an S-module isometry.*

Proof: First we assert (S,V,b) is S-anisotropic if and only if there is no

S-submodule $L \subset V$ which is a D-summand and for which $L \subset L^{\perp}$. Note that if $L = W \cap W^{\perp}$ then $L \subset L^{\perp}$. With $L_1 = \{v | v \in V, \ d \neq 0 \in D$ such that $dv \in L\}$ then $L_1^{\perp} = L^{\perp}$, $L_1 \subset L^{\perp}$ and L_1 is a D-summand.

If (S,V,b) is any orthogonal (skew-symmetric) representation, just take $L \subset V$ to be an S-submodule which is maximal with respect to $L \subset L^{\perp}$. Then $\langle S, L^{\perp}/L, \beta\rangle$ is clearly S-anisotropic, and by (I.1.12) is Witt equivalent to $\langle S,V,b\rangle$.

Let us not yet assume D is a field and still try to see how far we get on uniqueness. Thus (S,V,b) and (S,V_1,b_1) are assumed to be S-anistropic and $(S,V \oplus V_1, b \oplus (-b))$ is metabolic. Let $N \subset V \oplus V_1$ be a metabolizer. Note by the S-anistropic condition that

$$N \cap V = \{0\}$$

$$N \cap V_1 = \{0\} .$$

Then define an S-submodule $L \subset V$ by $v \in L$ if and only if there is a $v_1 \in V_1$ such that $(v,v_1) \in N$. Such a v_1 is unique so there is an S-module monomorphism $\Phi: L \to V_1$ with $\text{Graph}(\Phi) = N$. This is an isometry of $b|L$ into (V_1,b_1).

We assert that $L^{\perp} = \{0\}$. For if there were a $v \neq 0 \in L^{\perp}$ it would follow that $(v,0) \in N^{\perp} = N$ so that $v \in L$ and $\phi(v) = 0$. This means V/L is a finitely generated D-torsion module.

Obviously if D were a field, $L = V$ and it would similarly follow that $\Phi(L) = V_1$. This provides uniqueness. Caution, however, uniqueness fails completely for Dedekind domains. □

As the reader will realize there is within a Witt class a great deal of variation possible on the representation (S,V). Indeed there is little one can say about this that is invariant. Let us now say what is available.

If (S,L) is a representation of S then the dual representation (S,L^*) is given by

$$L^* = \mathrm{Hom}_D(L,D)$$

and $(s\varphi)(x) \equiv \varphi(\hat{s}x)$ for $s \in S$, $\varphi \in L^*$ and $x \in L$. Of course for an orthogonal (skew-symmetric) representation the adjoint provides an S-module isomorphism of

$$(S,V) \text{ with } (S,V^*) .$$

If $0 \to (S,L_1) \xrightarrow{\alpha} (S,L) \xrightarrow{\beta} (S,L_2) \to 0$ is a short exact sequence of representations then there is always a dual short exact sequence

$$0 \to (S,L_2^*) \xrightarrow{\beta^*} (S,L^*) \xrightarrow{\alpha^*} (S,L_1^*) \to 0$$

because all modules involved are D-finitely generated and projective. Thus if $GR(D,S)$ is the Grothendieck group of representations of S (short exact sequences are made to split) then on $GR(D,S)$ we receive an automorphism of period two induced by $(S,L) \to (S,L^*)$. With respect to this action of C_2 there is the cohomology group $H^2(C_2;GR(D,S))$. Also there are homomorphisms

$$W_0(D,S) \to H^2(C_2;GR(D,S))$$
$$W_2(D,S) \to H^2(C_2;GR(D,S)) .$$

This assigns to (S,V,b) the class of $(S,V) \simeq (S,V^*)$. If (S,V,b) is metabolic we have a short exact sequence

$$0 \to N \xrightarrow{i} V \xrightarrow{P} N^* \to 0$$

so that $V = N + N^*$ in $GR(D,S)$.

In a rough sense this generalizes rank modulo 2, the rank invariant associated with the Witt group of a field. In a few cases, due to work of Swan [Sw 2], this homomorphism can be understood. We shall take this up later, but we wished to record this invariant in our introductory section.

In the most important case of a group ring, $S = D(\pi)$, there is a $Z/4Z$ graded commutative ring structure on

$$W_*(D,\pi) = W_0(D,\pi) \oplus W_2(D,\pi) .$$

If say $\langle \pi, V, b\rangle$ and $\langle \pi, V_1, b_1\rangle$ are given then on $V \otimes_D V_1$ we want a D-valued innerproduct β for which

$$\pm \beta(v \otimes v_1, w \otimes w_1) = b(v,w) b_1(v_1,w_1) .$$

There is such an innerproduct. But let us look at the situation more carefully. If b, b_1 are both symmetric then so is β and we take the $+$ sign. If say b is symmetric and b_1 is skew-symmetric then β is skew-symmetric and again we take $+$. The third case occurs when the innerproducts b and b_1 are both skew-symmetric. In this case $\beta = -(b \otimes b_1)$ and β is symmetric. We use the minus sign to agree with topological considerations.

The action of π on $V \otimes_D V_1$ is $g(v \otimes w) = (gv \otimes gw)$ and π acts as a group of isometries with respect to β. That this yields a well-defined graded commutative ring structure on $W_*(D,\pi)$ we leave as an exercise to the reader. What is the multiplicative identity?. There is a commutative ring structure on $GR(D,\pi)$ and the involution induced by dualizing is an automorphism of this ring. Thus $H^2(C_2;GR(D,\pi))$ is a ring also and

$$W_*(D,\pi) \to H^2(C_2;GR(D,\pi))$$

is a ring homomorphism.

2. Commutative algebras over a field

The simplest case to handle is that of $D = K$, a field, and S a commutative linear algebra with involution over K. An immediate example of S would be the group algebra $K(\pi)$ of a finite abelian group.

Every Witt class contains an S-anisotropic representative which is unique up to an S-module isometry. Let us examine the structure of S-anistropics further.

(2.1) <u>Lemma</u>: <u>If</u> (S,V,b) <u>is</u> S-<u>anisotropic and</u> $W \subset V$ <u>is an</u> S-<u>submodule then</u> V <u>splits into an orthogonal direct sum</u> $V = W \oplus W^{\perp}$.

Proof: By the S-anisotropic condition $W \cap W^{\perp} = \{0\}$. Thus if (W, b_1) is the form on W induced from b we see $\text{Ad}: W \to \text{Hom}_K(W,K)$ is a monomorphism. Since K is a field, $\text{Ad}: W \simeq \text{Hom}_K(W,K)$. Now if $(W^{\perp}, b_2)$ is the form restricted to $W^{\perp}$ we have $\text{Ad}: W^{\perp} \simeq \text{Hom}_K(W^{\perp},K)$ also. An S-module projection $p: V \to W$ is given as follows. For $v \in V$ let $p(v) \in W$ be the unique element for which $b_1(p(v),w) = b(v,w)$ for all $w \in W$. Clearly $v - p(v) \in W^{\perp}$. □

(2.2) <u>Lemma</u>: <u>If</u> $W \subset V$ <u>is an</u> S-<u>submodule of an</u> S-<u>anisotropic then the ideal</u> $\text{Ann}(W) \subset S$ is involution invariant.

Proof: By Ann(W) we understand the ideal of $s \in S$ for which $sW = \{0\}$. If $s \in \text{Ann}(W)$ then for all x, y in W

$$b(x, \hat{s}y) = b(sx, y) = 0$$

so that $\hat{s}y \in W \cap W^{\perp} = \{0\}$. Thus $\hat{s} \in \text{Ann}(W)$ also.

Consider the family, I, of ideals $\mathcal{M} \subset S$ for which

1. $\hat{\mathcal{M}} = \mathcal{M}$
2. $\mathcal{M}$ is maximal
3. $S/\mathcal{M}$ is finite over K .

Let (S,V,b) be an S-anisotropic. For $\mathcal{M} \in I$ put

$$W(\mathcal{M}) = \{v \mid v \in V,\ sv = 0\ \forall\ s \in \mathcal{M}\} .$$

Then $W(\mathcal{M})$ is an S-submodule (possible 0) and in fact $W(\mathcal{M})$ is a linear space over the quotient field $S/\mathcal{M}$.

(2.3) Lemma: There are finitely many ideals $\mathcal{M}_1,\ldots,\mathcal{M}_n$ in I with $W(\mathcal{M}_i) \neq 0$ and (S,V,b) is the orthogonal direct sum $\sum_{j=1}^{n} (S,W(\mathcal{M}_j),b_j)$ where b_j is the restriction to $W(\mathcal{M}_j)$ of b.

Proof: Let us first note that if $\mathcal{M} \neq \mathcal{M}'$ then $W(\mathcal{M}') \subset W(\mathcal{M})^{\perp}$. Since $\mathcal{M} \neq \mathcal{M}'$ then by maximality we can find $s \in \mathcal{M}$, $s' \in \mathcal{M}'$ so that $s + s' = 1$. Then for $x \in W(\mathcal{M})$, $y \in W(\mathcal{M}')$, $b(x,y) = b((s+s')x,y) = b(sx,y) + b(s'x,y) = b(s'x,y) = b(x,\hat{s}'y) = 0$

Now there is a non-trivial irreducible S-submodule $L \subset V$. Of course L has the form $S/\mathcal{M}_1$ for some $\mathcal{M}_1 \in I$. Then $W(\mathcal{M}_1) \neq 0$ and $V = W(\mathcal{M}_1) \oplus W(\mathcal{M}_1)^{\perp}$. If $W(\mathcal{M}_1)^{\perp} \neq 0$ then there is an $\mathcal{M}_2$ with $0 \neq W(\mathcal{M}_2) \subset W(\mathcal{M}_1)^{\perp}$. Continuing we finally have

$$(S,V,b) = \sum_{j=1}^{n} (S,W(\mathcal{M}_j),b_j)$$

as required. The ideals $\mathcal{M}_1,\ldots,\mathcal{M}_n$ are unique. □

Notice that each quotient field $S/\mathcal{M}$ inherits an involution (possibly the identity). Thus $W_*(K,S/\mathcal{M})$ is defined for each $\mathcal{M} \in I$. Furthermore there is a homomorphism $W_*(K,S/\mathcal{M}) \to W_*(K,S)$ obtained by regarding a vector space over $S/\mathcal{M}$ as an S-module. We are suggesting

(2.4) Theorem: There is a canonical isomorphism

$$\sum_{I} W_*(K,S/\mathcal{M})) \simeq W_*(K,S) \quad .$$

Proof: Since every element in $W_*(K,S)$ has an S-anisotropic representative it surely follows from (I.2.3) that this is an epimorphism. To show it is a monomorphism we need the converse of (I.2.3).

(2.5) Lemma: Suppose for $1 \leq j \leq n$ we have non-trivial anisotropics $(S/\mathcal{M}_j,W_j,b_j)$, then $(S, \sum_{j=1}^{n} W_j,b)$ is S-anisotropic.

Proof: This is by induction on n. It is a tautology for $n=1$. Suppose there were an S-submodule $L \subset \sum_{j=1}^{n} W_j$ with $L \subset L^{\perp}$. There is an $s \in \mathcal{M}_n$ for which $s \notin \mathcal{M}_1 \cap \ldots \cap \mathcal{M}_{n-1}$. If $sL = 0$ then $L \subset W_n$, which contradicts W_n being anisotropic. If $sL \neq 0$ then $sL \subset \sum_{j=1}^{n-1} W_j$ which contradicts the inductive assumption. This also completes the proof of (I.2.4). □

Now to compute $W_*(K, S/\mathcal{M})$. We shall assume K is perfect so that $S/\mathcal{M}$ is a finite separable extension of K equipped with a Galois involution over K. This involution could be the identity and $S/\mathcal{M} = K$ is also permitted.

We want to discuss the Witt classification of $S/\mathcal{M}$-valued Hermitian innerproduct spaces over the field $S/\mathcal{M}$. For convenience let $E = S/\mathcal{M}$ and let $F \subset E$ be the fixed field of the involution. A Hermitian innerproduct space (V,h) is then a finite dimensional vector space over E together with an E-valued form for which

$$h(\alpha v + \beta v_1, w) = \alpha h(v,w) + \beta h(v_1, w)$$

$$h(w,v) = \hat{h}(v,w)$$

$$\mathrm{Ad}: V \simeq \mathrm{Hom}_E(V,E) .$$

The adjoint associates to $w \in V$ the homomorphism $v \to h(v,w)$. The linearity is in the first variable. By analogy with $W(E)$ there is the Witt group $\mathcal{H}(E) = \mathcal{H}_0(E)$ of Hermitian forms. If the involution is the identity then in fact $\mathcal{H}(E) = W(E)$.

We expect to show $\mathcal{H}(S/\mathcal{M}) \simeq W_0(K, S/\mathcal{M})$. We begin with some observations about $tr_{E/K}: E \to K$. If V is a linear space over E there is $\mathrm{Hom}_E(V,E)$ and $\mathrm{Hom}_K(V,K)$. Each is a vector space over E. If $\psi \in \mathrm{Hom}_E(V,E)$ then $(\alpha\psi)(v) \equiv \psi(\hat{\alpha} v) = \hat{\alpha}\psi(v)$. If $\varphi \in \mathrm{Hom}_K(V,K)$ then $(\alpha\varphi)(v) \equiv \varphi(\hat{\alpha} v)$. Here $\alpha \in E$. We introduce $\mathrm{Tr}: \mathrm{Hom}_E(V,E) \to \mathrm{Hom}_K(V,K)$ by

$$(\mathrm{Tr}\,\psi)(v) \equiv tr_{E/K}(\psi(v)) .$$

Remember $tr_{E/K}$ is K-linear. In fact Tr is E-linear. For $\alpha \in E$

$$\begin{aligned} Tr(\alpha\psi)(v) &= tr(\psi(\hat{\alpha}v)) \\ &= (Tr\ \psi)(\hat{\alpha}v) = (\alpha\, Tr\,\psi)(v) \ . \end{aligned}$$

(2.6) <u>Lemma</u>: <u>For every finite dimensional vector space over</u> E

$$Tr: Hom_E(V,E) \simeq Hom_K(V,K) \ .$$

Proof: Consider first $V = E$. Then $E \simeq Hom_E(E,E)$ by

$$\psi\alpha(\beta) = \beta\hat{\alpha} \ .$$

Since E is a separable extension $E \simeq Hom_K(E,K)$ by $\varphi_\alpha(\beta) = tr\ \beta\hat{\alpha}$. This is sufficient to establish the lemma. □

Suppose now we are given a Hermitian innerproduct space (V,h). There is

$$\begin{array}{ccc} V & \xrightarrow{Ad^h} & Hom_E(V,E) \\ & & \downarrow Tr \\ & & Hom_K(V,K) \end{array}$$

so we put $b(v,w) = Tr\, h(v,w)$. Then $Ad^b = Tr \circ Ad^h$. For $\alpha \in E$ $b(\alpha v,w) = tr(h(\alpha v,w))$ $= tr(h(v,\hat{\alpha}w)) = b(v,\hat{\alpha}w)$. Also $\beta(w,v) = tr(h(w,v)) = tr(\hat{h}(v,w)) = tr(h(v,w)) =$ $\beta(v,w)$. Remember the involution is a Galois automorphism over K.

We can reverse the process. Given $b(v,w)$ we proceed. With w fixed there is a unique $\psi_w \in Hom_E(V,E)$ with $tr\,\psi_w(v) \equiv b(v,w)$. Let $h(v,w) = \psi_w(v)$. This is linear in v. If $\alpha \in E$ then $b(\alpha v,w) = b(v,\hat{\alpha}w)$ so that $\psi_{\hat{\alpha}w}(v) = \psi_w(\alpha v) = \alpha\psi_w(v)$. This implies $h(v,\hat{\alpha}w) = h(\alpha v,w) = \alpha h(v,w)$. Finally $b(v,w) = b(w,v)$, and from this we deduce $h(v,w)$ is Hermitian as follows. For every $\alpha \in E$

$$\begin{aligned}&tr(h(\alpha v,w) - \hat{h}(\hat{\alpha}w,v))\\ &= tr(\alpha(h(v,w) - \hat{h}(w,v)))\\ &= b(\alpha v,w) - b(\hat{\alpha}w,v) = 0 .\end{aligned}$$

Since this is true for all $\alpha \in E$, $h(v,w) - \hat{h}(w,v) = 0$ as required.

The correspondence will send $\mathcal{H}(E)$ to $W_0(K,E)$. Suppose the Hermitian inner-product has a metabolizer $N \subset V$. This is an E subspace and $N = N^{\perp}$ with respect to h. With $v \in V$ assume $b(v,w) = 0$ for all $w \in N$. Since $tr(h(v,w)) = b(v,w)$ and

$$Tr: Hom_E(N,E) \simeq Hom_K(N,K)$$

it follows $h(v,w) = 0$ for all $w \in N$. Thus $v \in N$. This shows N is a metabolizer for (V,h). Combining with (I.2.4) we have as corollary

(2.7) <u>Corollary</u>: <u>If K is a perfect field then there is an isomorphism</u>

$$\sum_I \mathcal{H}(S/\mathcal{M}) \simeq W_0(K,S) . \quad \square$$

We shall discuss $W_2(K,S)$ later. Suppose now that $K = \mathbf{R}$, the real numbers. Each $S/\mathcal{M}$ would be an algebraic extension of $\mathbf{R}$. There only the two possibilities $S/\mathcal{M} = \mathbf{R}$ or $S/\mathcal{M} = C$. In the first case the induced involution is the identity and $\mathcal{H}(S/\mathcal{M}) = W(\mathbf{R}) \simeq Z$. The last is an isomorphism given by signature. For C if the induced involution is complex conjugation then $\mathcal{H}(C) \simeq Z$ again by signature, but if the involution is the identity then $W(C) \simeq Z/2Z$ with rank mod 2 as the isomorphism.

If K is a finite field then $\mathcal{H}(S/\mathcal{M}) \sim Z/2Z$ if the induced involution is non-trivial and $\mathcal{H}(S/\mathcal{M}) \simeq W(S/\mathcal{M})$ if the involution is the identity ([M-H]).

For a cyclic group of order n we might consider the group algebra $K(C_n)$. If T is a generator of C_n then the involution is $T \longrightarrow T^{-1}$. We think of $K(C_n)$ as the quotient $K(t)/(t^n-1)$. To every irreducible monic polynomial $\phi(t)$ which divides t^n-1 and for which

$$\frac{t^{\deg \phi(x)}}{a_0} \phi(t^{-1}) = \phi(t)$$

where a_0 is the constant term there corresponds a maximal ideal $\mathcal{M} \subset K(C_n)$ for which $\hat{\mathcal{M}} = \mathcal{M}$. Actually $\mathcal{M}$ is the principal ideal generated by $\phi(t) \in K(C_n)$. This describes all the ideals in the class I. Obviously $(t-1)$ is always one such member. Thus

$$W_0(K, C_n) = \sum \mathcal{H}(K[t]/\phi(t)))$$

where the sum is taken over all monic irreducible factors $\phi(t)$ of t^n-1 that satisfy $\frac{t^{\deg \phi}}{a_0} \phi(t^{-1}) = \phi(t)$. The contribution from $t-1$ is just $W(K)$.

For example, if $\mathbf{R} = K$ we put $\lambda = \exp 2\pi i/n$ so that we get

$$t^n-1 = (t-1) \prod_{j=1}^{(n-1)/2} (t^2 - (\lambda^j + \lambda^{-j})t + 1) \quad \text{if } n \text{ is odd or}$$

$$t^n-1 = (t-1)(t+1) \sum_{j=1}^{(n-2)/2} (t^2 - (\lambda^j + \lambda^{-j})t + 1)$$

if n is even. This tells us $W_0(\mathbf{R}, C_n)$ is a free abelian group of rank $(n+1)/2$ if n is odd and of rank $(n+2)/2$ if n is even. The point is that $t-1$ or $t+1$ give $\mathbf{R}[t]/(\phi(t)) = \mathbf{R}$. But $\mathbf{R}[t]/(t^2 - (\lambda^j + \lambda^{-j})t + 1) = C$ with complex conjugation as the involution. Later we shall look at other ways of viewing this.

If $K = Q$ then t^n-1 is a product of the cyclotomic polynomials $\phi_m(t)$ for each m/n. We agree $\phi_1(t) = t-1$ and $\phi_2(t) = t+1$. Each $\phi_m(t)$ is irreducible and satisfies the required symmetry. Thus $W_0(Q, C_n) \simeq W(Q) \oplus \sum \mathcal{H}(Q(\lambda_d))$ or $W(Q) \oplus W(Q) \oplus \sum \mathcal{H}(Q(\lambda_d))$. Here $Q(\lambda_d)$ is the cyclotomic number field arising from the adjunction of a primitive d^{th} root of unity, $d \geq 2$. The involution on $Q(\lambda_d)$ is complex conjugation.

Due to Landherr's work ([L]) it is possible to determine $\mathcal{H}(Q(\lambda_d))$. In a future section we shall take this up. The reader may wish to consider π a finite abelian group and run a similar analysis.

Actually to handle $W_2(K,S)$ it is simplest to assume $ch(K) \neq 2$. With this assumption symplectic is equivalent to skew-symmetric. Then we can consider skew-Hermitian innerproducts. That is, $h(w,v) = -\hat{h}(v,w)$. They have a Witt group $\mathcal{H}_2(S/\mathcal{M}) \simeq W_2(K,S/\mathcal{M})$. If the involution on $S/\mathcal{M}$ is trivial then $\mathcal{H}_2(S/\mathcal{M}) = W_2(S/\mathcal{M}) \simeq W_2(K,S/\mathcal{M}) = 0$. If the involution on $S/\mathcal{M}$ is non-trivial then $\mathcal{H}_2(S/\mathcal{M}) \simeq \mathcal{H}_0(S/\mathcal{M}) \simeq W_2(K,S/\mathcal{M}) \simeq W_0(K,S/\mathcal{M})$. To see this choose any element $\gamma \in S/\mathcal{M}$ with $\hat{\gamma} = -\gamma \neq 0$. If $h(v,w)$ is skew-Hermitian then $\gamma h(v,w)$ is Hermitian.

As a consequence if $K = \mathbb{R}$ then $W_2(\mathbb{R},S)$ is a free abelian group. Only those maximal ideals for which $S/\mathcal{M}$ has a non-trivial involution make a contribution. Thus $W_2(R,C_n)$ has rank $(n-1)/2$ if n is odd or $(n-2)/2$ if n is even.

3. Torsion forms

In this section D is again a Dedekind domain with K as its field of fractions. At the outset we again allow S to be non-commutative. A torsion form is a pair (A,β) in which A is a finitely generated torsion D-module and $\beta: A \times A \to K/D$ is a D-bilinear, symmetric (skew-symmetric) form with values in K/D such that $Ad: A \simeq Hom_D(A,K/D)$. The adjoint is defined as usual. We are then interested in triples (S,A,β) in which (S,A) is an S-module structure on A extending the finitely generated torsion D-module structure and

$$\beta(sa,a') \equiv \beta(a,\hat{s}a') .$$

We shall use the fact that K/D is D-injective to prove some elementary facts. For any S submodule $B \subset A$ let

$$B^{\perp} = \{a | a \in A, \beta(a,b) = 0 \ \forall \ b \in B\} .$$

Of course $B^{\perp}$ is again an S-submodule. Compare the following with (I.1.3).

(3.1) <u>Lemma</u>: <u>For any S-submodule</u> $B \subset A$, $(B^{\perp})^{\perp} = B$.

Proof: As before we ignore the S-module structure. We define

$P: A \to \mathrm{Hom}_D(B,K/D)$ by $P(a)(b) = \beta(a,b)$. Because K/D is injective this is an epimorphism so we have

$$0 \to B^{\perp} \xrightarrow{i} A \xrightarrow{P} \mathrm{Hom}_D(B,K/D) \to 0 .$$

Appealing again to the injectiveness of K/D there is a dual exact sequence

$$0 \to \mathrm{Hom}_D(\mathrm{Hom}_D(B,K/D),K/D) \xrightarrow{P^*} \mathrm{Hom}_D(V,K/D) \xrightarrow{i^*} \mathrm{Hom}_D(B^{\perp},K/D) \to 0 .$$

This is to be compared with

$$0 \to (B^{\perp})^{\perp} \longrightarrow V \xrightarrow{\tilde{P}} \mathrm{Hom}_D(B^{\perp},K/D) \to 0$$

just as in the proof of (I.1.3). The proof of (3.1) is completed by observing that the canonical homomorphism $B \to \mathrm{Hom}_D(\mathrm{Hom}_D(B,K/D),K/D)$ is again an isomorphism. □

(3.2) Lemma: If B and B_1 are S-submodules of (S,A,β) then

$$(B + B_1)^{\perp} = B^{\perp} \cap B_1^{\perp}$$

$$(B \cap B_1)^{\perp} = B^{\perp} + B_1^{\perp} .$$

Proof: The first statement is elementary. For the second use (I.3.1) to write

$$(B^{\perp} + B_1^{\perp})^{\perp} = (B^{\perp})^{\perp} \cap (B_1^{\perp})^{\perp} = B \cap B_1$$

and therefore by (I.3.1) again

$$(B \cap B_1)^{\perp} = B^{\perp} + B_1^{\perp} . \quad \square$$

It is interesting to note that if (S,A,β) is metabolic and if $L \subset A$ is an S-submodule with $L \subset L^{\perp}$ then $L + L^{\perp} \cap N$ is a metabolizer if N is a metabolizer

Simply consider

$$(L + L^{\perp} \cap N)^{\perp} = L^{\perp} \cap (L^{\perp} \cap N)^{\perp} = L^{\perp} \cap (L + N)$$
$$= L + L^{\perp} \cap N .$$

The reader may go on to introduce the Witt groups $W_0(K/D,S)$ and $W_2(K/D,S)$. We wish to look further at K/D.

If $P \subset D$ is a prime ideal let $\mathcal{O}(\mathcal{P}) \subset K$ be the local ring of integers associated with $\mathcal{P}$. Since $D \subset \mathcal{O}(\mathcal{P})$ there is a D-module epimorphism $K/D \to K/\mathcal{O}(\mathcal{P}) \to 0$.

(3.3) Lemma: There results a D-module isomorphism $K/D \simeq \sum K/\mathcal{O}(\mathcal{P})$.

Proof: Since $D = \cap\, \mathcal{O}(\mathcal{P})$ this is surely a monomorphism. Then from the Strong Approximation Theorem ([OM.]) we assert that given primes $\mathcal{P}_1,\ldots,\mathcal{P}_n$ and field elements $k_1,\ldots,k_n$ then there is a $k \in K$ such that $k \in \mathcal{O}(\mathcal{P})$ if $\mathcal{P} \notin \{\mathcal{P}_1,\ldots,\mathcal{P}_n\}$ and $k-k_j \in \mathcal{O}(\mathcal{P}_j)$, $1 \leq j \leq n$. Surely then we have an epimorphism. □

If Π is a local uniformizer on $\mathcal{O}(\mathcal{P})$ (i.e, a generator for the unique maximal ideal in $\mathcal{O}(\mathcal{P})$) then the residue field $D/\mathcal{P}$ embeds in K/D by a composition $D/\mathcal{P} \to K/\mathcal{O}(\mathcal{P}) \to K/D$. The first homomorphism is $d \to c\ell(d\pi^{-1},\mathcal{O}(\mathcal{P}))$ by which we mean the class of $d\pi^{-1}$ in $K/\mathcal{O}(\mathcal{P})$. We get the same effect by projecting $c\ell(d\pi^{-1}) \in K/D$ into $K/\mathcal{O}(\mathcal{P})$. This is a D-module embedding. Thinking of $K/\mathcal{O}(\mathcal{P})$ as an $\mathcal{O}(\mathcal{P})$-module then $D/\mathcal{P}$ is $\mathcal{O}(\mathcal{P})/(\pi)$. It is assumed that a choice of local uniformizers has been made.

Now let us assume (S,A,β) is S-anisotropic. If $\mathcal{P} \subset D$ is a prime ideal let $A(\mathcal{P}) = \{v \mid v \in A,\ \mathcal{P}v = 0\}$. Suppose this is non-trivial. It is certainly an S-submodule and $A = A(\mathcal{P}) \oplus A(\mathcal{P})^{\perp}$. Thus $\beta \mid A(\mathcal{P})$ is still an innerproduct. If $\beta_{\mathcal{P}} = \beta/A(\mathcal{P})$ then the value $\beta_{\mathcal{P}}(a,a')$ lies exactly in $K/\mathcal{O}(\mathcal{P}) \subset K/D$ and in fact in the image of $D/\mathcal{P} \to K/\mathcal{O}(\mathcal{P}) \subset K/D$. Thus since $A(\mathcal{P})$ is a vector space over $D/\mathcal{P}$ it is correct to think of $(A(\mathcal{P}),\beta_{\mathcal{P}})$ as an innerproduct with values in $D/\mathcal{P}$. Furthermore $\mathcal{P} \subset D$ generates a 2-sided ideal $\mathcal{P}S \subset S$ so there is $S(\mathcal{P}) = S/\mathcal{P}S$ which is a linear algebra with involution over $D/\mathcal{P}$. Thus $(S(\mathcal{P}),A(\mathcal{P}),\beta_{\mathcal{P}})$ defines an element in

$W_*(D/\mathcal{P}, S(\mathcal{P}))$. By analogy with section 2 we get

(3.4) Theorem: There is an isomorphism

$$\sum_{\mathcal{P} \subset D} W_*(D/\mathcal{P}, S(\mathcal{P})) \simeq W_*(K/D, S) \ . \ \square$$

For $D = Z$, $S = Z(\pi)$ this says

$$W_*(Q/Z, \pi) \simeq \sum W_*(F_p, \pi)$$

where F_p is the field of integers mod p. To go further we must suppose S is commutative. For $\mathcal{P} \subset D$ let $I(\mathcal{P})$ be the family of ideals $\mathcal{M} \subset S$ with

1. $\mathcal{M} \supset \mathcal{P}S$
2. $\hat{\mathcal{M}} = \mathcal{M}$
3. $\mathcal{M}$ is maximal
4. $S/\mathcal{M}$ is a finite extension of $D/\mathcal{P}$.

Note that 1 is equivalent to $\mathcal{P} = \mathcal{M} \cap D$. This family could be empty. If so $W_*(D/\mathcal{P}, S(\mathcal{P})) = 0$. Now we say

(3.5) Theorem: There is a canonical isomorphism

$$\sum_{I(\mathcal{P})} \mathcal{H}_0(S/\mathcal{M}) \simeq W_0(D/\mathcal{P}, S(\mathcal{P})) \ .$$

In cases of interest $D/\mathcal{P}$ and hence $S/\mathcal{M}$ will be finite fields. We go, as in section 2, from $\mathcal{H}_0(S/\mathcal{M})$ to $W_0(D/\mathcal{P}, S(\mathcal{P}))$ by using trace of $S/\mathcal{M}$ over $D/\mathcal{P}$ (assumed perfect) and by the change of rings $S(\mathcal{P}) \to S/\mathcal{M} \to 0$. We can handle $W_2(D/\mathcal{P}, S(\mathcal{P}))$ by agreeing to replace symplectic by skew-symmetric. Correspondingly $\mathcal{H}_2(S/\mathcal{M})$ replaces $\mathcal{H}_0(S/\mathcal{M})$.

For $D = Z$ and $S = Z(C_n)$ let us follow out the prescription for computing $W_*(Q/Z, C_n)$. First $W_*(Q/Z, C_n) \simeq \sum W_*(F_p, C_n)$. For each rational prime we would factor t^n-1 in $F_p[t]$. For each irreducible factor $\phi(t)$ that satisfies

$$\frac{t^{\deg \phi}}{a_0}\phi(t^{-1}) = \phi(t)$$

we would get a quotient field $F_p[t]/(\phi(t))$ with involution. If the involution is non-trivial or $p=2$ $\mathcal{H}(F_p[t]/(\phi(t)) \simeq Z/2Z$. If it is the identity and p is odd, then $\mathcal{H}(F_p[t]/(\phi(t)) = W(F_p[t]/(\phi(t)) \simeq Z/4Z$ or $Z/2Z \oplus Z/2Z$.

4. An exact sequence

For a Dedekind domain with K as its field of fractions, Knebusch ([M-H]) introduced a sequence

$$0 \to W(D) \to W(K) \xrightarrow{\partial} \sum_{\mathcal{P}\subset D} W(D/\mathcal{P})$$

relating Witt groups over D, K and the residue fields of D. We would like something similar for $W_*(D,S)$. For this we shall assume S is finitely generated as a D-module. We shall also need $\mathcal{S} = S \otimes_D K$ which will be a linear algebra over K with involution. First we have $0 \to W_*(D,S) \to W_*(K,\mathcal{S})$. Given (S,V,b) we form $(\mathcal{S},V\otimes_D K,B)$ by tensoring (V,b) with the innerproduct on K given by $(k,k') \to kk'$. This preserves Witt classes and in fact induces a monomorphism for if $N \subset V\otimes_D K$ is a metabolizer then $N \cap V$ is immediately a metabolizer for (S,V,b).

We are concerned with

$$\partial: \ W_*(K,\mathcal{S}) \to W_*(K/D,S) \ .$$

Suppose we are given $(\mathcal{S},V,B)$ over K. We assert, and shall prove at the end of this section, that there is a lattice $L \subset V$ such that

1) L is a finitely generated D-submodule of V
2) L is an S-submodule of V
3) $L \otimes_D K = V$

4) $B(v,w) \in D$ for all v,w in L.

Define $L^{\#} \subset V$ by

$$L^{\#} = \{v \mid v \in V, B(v,w) \in D \ \forall \ w \in L\} .$$

Then $L \subset L^{\#}$ and in fact $L^{\#}$ can be naturally identified with $\text{Hom}_D(L,D)$. Furthermore $L^{\#}$ is an S-submodule. Thus $L^{\#}$ is a lattice also and $(L^{\#})^{\#} = L$. Now $L^{\#}/L$ is a finitely generated torsion D-module and it receives a natural S-module structure. We define $\beta: L^{\#}/L \times L^{\#}/L \to K/D$ by

$$\beta(v+L, w+L) = cl(B(v,w)) \in K/D$$

for all v, w in $L^{\#}$. This is seen to be well defined and correctly related to the S-module structure.

We want to see $\text{Ad}_\beta: L^{\#}/L \simeq \text{Hom}_D(L^{\#}/L, K/D)$. Now V/L is a torsion D-module and $L^{\#}/L \subset V/L$. Given a D-module homomorphism $\varphi: L^{\#}/L \to K/D$ then since $L^{\#}$ is a projective D-module there is a commutative diagram of D-module homomorphisms

$$
\begin{array}{ccc}
L^{\#} & \xrightarrow{\ \psi\ } & K \\
\downarrow & & \downarrow \\
L^{\#}/L & \xrightarrow{\ \varphi\ } & K/D
\end{array}
$$

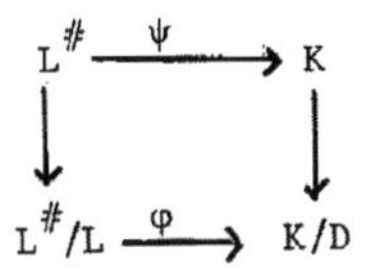

Then ψ extends uniquely to a K-linear $\psi: V \to K$. Hence there is a unique $v \in V$ with $B(w,v) = \psi(w)$ for all $w \in V$. But if $w \in L$, $\psi(w) \in D$ so $v \in L^{\#}$. This shows the adjoint is an epimorphism. Since $(L^{\#})^{\#} = L$ it is also a monomorphism.

Suppose next that $(S, L^{\#}/L, \beta)$ is metabolic. Choose a metabolizer $\tilde{N} \subset L^{\#}/L$. Let $W \subset L^{\#}$ be the pull back of $\tilde{N}$ under $L^{\#} \to L^{\#}/L$. This is an S-submodule. We need to show $W^{\#} = W$. Clearly $W \subset W^{\#}$. If $v \in W^{\#}$ then $v \in L^{\#}$ so the image of v under $L^{\#} \to L^{\#}/L$ lies in $\tilde{N}$ and so by definition $v \in W^{\#}$. Hence we have $\langle S, W, b \rangle \in W_*(D,S)$ whose image in $W_*(K, \mathcal{S})$ is $\langle \mathcal{S}, V, B \rangle$.

We shall have our exact sequence if we can show ∂ is well defined. Suppose $(\mathcal{S},V,B)$ has a metabolizer M. Let $N=M\cap L^{\#}$. Suppose $v\in L^{\#}$ has $B(v,w)\in D$ for all $w\in N$. Since $N=N^{\perp}$ in $L^{\#}$ it is a D-summand of $L^{\#}$ thus the D-module homomorphism $N\to D$ given by $v\to B(w,v)$ can be extended to $\varphi\colon L^{\#}\to D$. Now $(L^{\#})^{\#}=L$ so there is a $v_0\in L$ with $B(w,v-v_0)=0$ for all $w\in N$. Clearly $v-v_0\in N$. Thus putting $\tilde{N}\subset L^{\#}/L$ equal to the image of N we find $\tilde{N}$ is a metabolizer for $(S,L^{\#}/L,\beta)$. This shows ∂ is well defined for if $(\mathcal{S},V,B)\sim(\mathcal{S},V_1,B_1)$ we choose lattices $L\subset V$, $L_1\subset V_1$ and $L\oplus L_1$ is a lattice for $(\mathcal{S},V\oplus V_1,B\oplus(-B_1))$. Applying the foregoing $\langle S,L^{\#}/L_1,\beta\rangle=\langle S,L_1^{\#}/L_1,\beta_1\rangle\in W_*(K/D,S)$.

(4.1) __Theorem:__ __There is an exact sequence__

$$0\to W_*(D,S)\to W_*(K,\mathcal{S})\xrightarrow{\partial} W_*(K/D,S)\ . \quad \square$$

We have no idea as to how to extend this sequence to the right. In (I.3.4) we showed how

$$\sum W_*(D/\mathcal{P},S(\mathcal{P}))\simeq W_*(K/D,S)\ .$$

In this manner we have for each prime in D

$$\partial_{\mathcal{P}}\colon W_*(K,\mathcal{S})\to W_*(D/\mathcal{P},S(\mathcal{P}))\ .$$

In fact if D is replaced by $\mathcal{O}(\mathcal{P})$ and S by $\mathcal{S}(\mathcal{P})=S\otimes_{\mathfrak{n}}\mathcal{O}(\mathcal{P})$ we get

$$0\to W_*(\mathcal{O}(\mathcal{P}),\mathcal{S}(\mathcal{P}))\to W_*(K,\mathcal{S})\xrightarrow{\partial_{\mathcal{P}}} W_*(D/\mathcal{P},S(\mathcal{P}))\ .$$

This is the localized version of (4.1).

We promised

(4.2) __Lemma:__ __There is a lattice__ $L\subset V$ __for any__ $(\mathcal{S},V,B)$.

Proof: Let $s_1,\ldots,s_k$ be a finite set of elements which generate S as a D-module. Let $e_1,\ldots,e_n$ be a basis for V over K. We can choose $d \neq 0$ in D so that all

$$B(ds_i e_j, ds_k e_\ell) = d^2 B(s_i e_j, s_k e_\ell) \in D .$$

Take L to be the D-submodule generated by all $ds_i e_j$. This also is an S-submodule and has the required properties.

5. The Atiyah-Bott homomorphism

This section is not as general as the four previous. For a finite group we have introduced $W_*(D,\pi)$. If D had an involution, $d \to \bar{d}$, we could also introduce $\mathcal{H}_*(D,\pi) = \mathcal{H}_0(D,\pi) \oplus \mathcal{H}_2(D,\pi)$ by using Hermitian (skew-Hermitian) innerproduct spaces over D together with an action of π as a group of isometries. In this section we specialize to $D = C$ with complex conjugation and π a finite abelian group. Our basic task is to show

(5.1) Theorem: There is a canonical isomorphism, for π abelian,

$$\mathcal{H}_*(C,\pi) \simeq Z(C_2)(\mathrm{Hom}(\pi, C^*)) .$$

We mean $Z(C_2)$ is the group ring over Z of the cyclic group of order 2 and we form the group ring of the dual group over this ring. The isomorphism is the multisignature.

Suppose (π, V, h) is a Hermitian (skew-Hermitian) representation of π. For each homomorphism $\chi: \pi \to C^*$ let V_χ be the π-invariant subspace

$$V_\chi = \{v \mid v \in V, gv = \chi(g)v, \quad \text{all} \quad g \in \pi\} .$$

There is a direct sum decomposition $V = \sum_{\chi \in \mathrm{Hom}(\pi, C^*)} V_\chi$. In fact for each χ the projection $P_\chi: V \to V_\chi$ is

$$P_\chi(v) = \frac{1}{\#(\pi)} \sum_{h\in\pi} h\ \chi(h^{-1})v$$

Furthermore, if $\chi \neq \chi_1$ then $V_\chi^\perp \supset V_{\chi_1}$ for if $v \in V_\chi$, $v_1 \in V_{\chi_1}$ then for all $g \in \pi$

$$\begin{aligned} h(v,v_1) &= h(gv,gv_1) = h(\chi(g)v,\chi_1(g)v_1) \\ &= \chi(g)\overline{\chi_1(g)}h(v,v_1) \ . \end{aligned}$$

Since $\chi(g)\overline{\chi_1(g)} \neq 1$ for some $g \in \pi$ we find $h(v,v_1) = 0$. Thus $V = \sum V_\chi$ is actually an orthogonal direct sum decomposition and (V_χ, h_χ) is still a Hermitian (skew-Hermitian) innerproduct space where h_χ represents the restriction to V_χ of h. The next is left to the reader.

(5.2) Lemma: The projection $P_\chi : V \to V_\chi$ is self-adjoint in the sense $h(v,P_\chi(w)) = h(P_\chi(v),w)$. $\square$

Using this we suppose $N \subset V$ is a π-invariant metabolizer. We claim $N_\chi \subset V_\chi$ is a metabolizer for (V_χ, h_χ). Suppose $v \in V_\chi$ and $h(v,P_\chi(w)) = 0$ for all $w \in N$. Then $h(P_\chi(v),w) \equiv 0$ for all $w \in N$. Hence $P_\chi(v) = v \in N$ so $v \in V_\chi \cap N = N_\chi$.

Thus the element $\langle \pi,V,h\rangle$ in $\mathcal{H}_0(C,\pi)$ (or in $\mathcal{H}_2(C,\pi)$) is determined by the set $\{\langle V_\chi, h_\chi\rangle\}$ in $\mathcal{H}_0(C)(\mathcal{H}_2(C))$. This means

$$\mathcal{H}_*(C,\pi) \simeq \mathcal{H}_*(C)(\pi^*) \ ,$$

the group ring over $\mathcal{H}_*(C)$ of the dual group $\pi^* = \mathrm{Hom}(\pi, C^*)$. Remember always our multiplication convention that if (V,h) and (V_1,h_1) are both skew-Hermitian then we use $(V \otimes_C V_1, -(h \otimes h_1))$. If $\langle i\rangle \in \mathcal{H}_2(C)$ is the rank 1 skew-form on C given by $h(z,z_1) = i\, z\, \bar{z}$, then according to our product rule

$$\langle i\rangle^2 = \langle -i\cdot i\rangle = \langle 1\rangle \ .$$

Thus multiplication by $\langle i \rangle$ yields an isomorphism

$$\mathcal{H}_2(C) \simeq \mathcal{H}(C) .$$

Thus $\mathcal{H}_*(C) = \mathcal{H}(C)(C_2)$. Via signature, $\operatorname{sgn} \mathcal{H}(C) \simeq Z$. Hence we arrive at $\mathcal{H}_*(C,\pi) \simeq (Z(C_2))(\pi^*)$. Explicitly, given $(\langle \pi,V,h\rangle, \langle \pi,V',h^1\rangle) \in \mathcal{H}(C,\pi) \oplus \mathcal{H}_2(C,\pi)$ we take this to

$$\sum_{\chi} (\operatorname{sgn}(V_\chi, h_\chi)I + \operatorname{sgn}(V', -ih^1_\chi)T)\chi$$

where $T \in C_2$ has $T^2 = I$.

Actually the Atiyah-Bott homomorphism

$$ab\colon W_*(Z,\pi) \to Z(C_2)(\pi^*)$$

is the ring homomorphism that arises from $\langle \pi,V,b\rangle \to \langle \pi, V \otimes_Z C, h\rangle$ where $h(v \otimes z, v_1 \otimes z_1) = b(v,v_1)z\bar{z}$. In case π is an abelian p-group this isomorphically embeds $W_*(Z,\pi)$ into $Z(C_2)(\pi^*)$. This will be seen much later.

Now associated with every group ring there is an augmentation and so we have

$$\mathcal{H}_*(C,\pi) \simeq Z(C_2)(\pi^*) \xrightarrow{\ \varepsilon\ } Z(C_2) .$$

For a pair $(\langle \pi,V,h\rangle, \langle \pi,V',h'\rangle)$ this would yield $\operatorname{sgn}(V,h)I + \operatorname{sgn}(V',-ih')T$. But one point had better be clarified.

Consider $W_*(\mathbb{R},\pi) \to \mathcal{H}_*(C,\pi)$ given by tensoring with C. This will produce a Hermitian (skew-Hermitian) (π,V,h) for which V is equipped with an involution $v \to \hat{v}$ which is π-equivariant and for which

$$\widehat{zv} = \bar{z}\,\hat{v}, \quad z \in C$$

$$h(\hat{v},\hat{v}_1) = \overline{h(v,v_1)} = \pm\, h(v_1,v) .$$

Thus we find (V_χ, h_χ) is conjugate equivalent to $(V_{\bar\chi}, h_{\bar\chi})$. In the Hermitian case $\operatorname{sgn}(V_\chi, h_\chi) = \operatorname{sgn}(V_{\bar\chi}, h_{\bar\chi})$. In the skew-Hermitian case, however,

$$\operatorname{sgn}(V_\chi, -ih_\chi) = -\operatorname{sgn}(V_{\bar\chi}, -ih_{\bar\chi}) .$$

This does mean that

$$W_2(\mathbb{R},\pi) \to \mathcal{H}_2(C,\pi) \xrightarrow{\varepsilon} Z(C_2)$$

is trivial (as it had better be!). Further, $W_0(\mathbb{R},\pi) \to \mathcal{H}_0(C,\pi) \xrightarrow{\varepsilon} Z$ is ordinary signature.

Chapter II

Smooth actions of C_{p^n} on even dimensional manifolds

Suppose p is an odd prime and T is an orientation preserving diffeomorphism of a closed, compact manifold M^{2m} of period p^n, $n > 0$. Let $W(F_p)$ be the Witt group of orthogonal forms over the field with p elements. In this chapter, we define an invariant $q(T,M) \in W(F_p)$ using the C_{p^n}-module structure on $H^*(M;Z)$ determined by T^* and the C_{p^n}-invariant nonsingular pairing on $H^m(M;Z)/TOR$. $q(T,M)$ is a cobordism invariant vanishing on fixed point free actions of C_{p^n} and can therefore be computed in terms of the fixed set, F, of T in M and the action of C_{p^n} in the normal tube about F. Theorem 4.1 gives the general formula for the local computation for all odd primes and we apply it to establish a mod 4 relationship between the signature of M and the signature of F and a mod 8 computation of the signature of the quotient space M/T. See [AHV1], [AHV2] for an earlier version involving C_p, p an odd prime.

1. Algebraic Preliminaries

Let p be an odd prime and $n > 0$ an integer. The involution on $Z[C_{p^n}]$ generated by sending $g \longrightarrow g^{-1}$ for all $g \in C_{p^n}$ is denoted $a \quad \hat{a}$. Let

$\Sigma = 1 + T + \ldots + T^{p^n - 1}$ and $\Delta = T^{\frac{p^n-1}{2}} - T^{\frac{p^n+1}{2}}$. Notice that $\hat{\Sigma} = \Sigma$ and $\hat{\Delta} = -\Delta$.

(1.1) Lemma: There is a unique $\theta \in Z[C_{p^n}]$ for which $\Delta\theta = p^n\Sigma$ and $\hat{\theta} = -\theta$.

Proof: First if $\varepsilon\colon Z[C_{p^n}] \to Z$ is augmentation then for any $X \in Z[C_{p^n}]$, $X\cdot\Sigma = \varepsilon(X)\Sigma$ because $T\Sigma = \Sigma$. Thus $\Sigma^2 = p^n\Sigma$ and so $\Sigma(p^n - \Sigma) = 0$. Since $Z[C_{p^n}]$ is a free module there is an X so that $\Delta X = p^n - \Sigma$. Since $\hat{\Delta} = -\Delta$, $\Delta(X + \hat{X}) = 0$ so $X + \hat{X} = m\Sigma$ for some integer m. But $\varepsilon(X) = \varepsilon(\hat{X})$ and $p^n = \varepsilon(\Sigma)$ is odd so $m = 2k$. Let $\theta = X - k\Sigma$. We leave it to the reader to show that θ satisfies the conclusion of the lemma and is unique. $\square$

Call a bilinear pairing $\varphi\colon A \times B \to Z$ of free abelian groups nonsingular if the adjoint map $\hat{\varphi}\colon A \to \mathrm{Hom}(B;Z)$ is an isomorphism. If A and B are C_{p^n}-modules this pairing is C_{p^n}-invariant if $\varphi(ga,gb) = \varphi(a,b)$ for all $g \in C_{p^n}$. For $x \in Z[C_{p^n}]$, $\varphi(xa,b) = \varphi(a,\hat{x}b)$. For an endomorphism $\alpha\colon A \to A$, let $A^\alpha = \ker\alpha$ and $_\alpha A = \mathrm{im}\,\alpha$. Then $H^1(C_{p^n};A) = A^\Sigma/_\Delta A$ and $H^2(C_{p^n};A) = A^\Delta/_\Sigma A$.

(1.2) Lemma: If A and B are free abelian C_{p^n}-modules and $\varphi\colon A \times B \to Z$ is a C_{p^n}-invariant nonsingular pairing then φ induces nonsingular finite group pairings

$$\varphi^1\colon H^1(C_{p^n};A) \times H^1(C_{p^n};B) \to Q/Z$$

and

$$\varphi^2\colon H^2(C_{p^n};A) \times H^2(C_{p^n};B) \to Q/Z$$

by

$$\varphi^1([x],[y]) = \frac{1}{p^n}\varphi(x,\theta y) \bmod Z$$

$$\varphi^2([x],[y]) = \frac{1}{p^n}\varphi(x,y) \bmod Z\ .$$

Proof: There is a well-defined pairing

$$A^{\Sigma}/\Delta A \times B^{\Sigma}/\Delta B \longrightarrow Q/Z \quad \text{because}$$

$$\begin{aligned}\varphi^1(a,\Delta b) &\equiv \frac{1}{p^n}\varphi(a,\Delta\theta b) \mod Z\\ &\equiv -\frac{1}{p^n}\varphi(a,\Sigma b) \mod Z\\ &\equiv -\frac{1}{p^n}\varphi(\Sigma a,b) \mod Z\\ &\equiv 0 \mod Z\ .\end{aligned}$$

We need to show that given a non-zero class $[a] \in H^1(C_{p^n};A)$, then there is a b] $\in H^1(C_{p^n};B)$ such that $\varphi^1([a],[b]) \neq 0$. Suppose that $^1([a],[b]) = 0$ for all $[b]$. Since φ is nonsingular there is an $x \in A$ such hat $\varphi(x,b) = \frac{1}{p^n}\varphi(a,\theta b)$ for all $b \in B^{\Sigma}$. This implies that $\varphi(x,p^n b) = \varphi(x,\Delta\theta b) =$ $(-\Delta x,\theta b)$. Hence $-\Delta x$ and a agree on $\theta(B^{\Sigma})$, but $\theta\colon B^{\Sigma} \to B^{\Sigma}$ is injective so $+\Delta x$ annihilates B^{Σ}. The annihilator of B^{Σ} is ${}_{\Sigma}A$ and since $\Sigma(a+\Delta x) = 0$ his implies $a+\Delta x \in {}_{\Sigma}A \cap A^{\Sigma} = \{0\}$. Therefore $[a] = 0$. This shows φ^1 is non- ingular.

Also notice that $\varphi^1(a,b) = \frac{1}{p^n}\varphi(a,\theta b) = -\frac{1}{p^n}\varphi(\theta a,b)$. A similar argument hows that φ^2 is nonsingular. Notice that if A and B are π-modules for a inite group π and φ is π-invariant with $T \in \text{Center}(\pi)$, then π/C_{p^n} acts on $^*(C_{p^n};A)$ and $H^*(C_{p^n};B)$ and the finite group pairings are left invariant by this ction. □

(1.3) Corollary: *If A is a free abelian C_{p^n}-module and $\varphi\colon A \times A \to \mathbb{Z}$ is a $_{p^n}$-invariant symmetric (skew-symmetric) nonsingular bilinear form then $\varphi^2(\varphi^1)$ is nonsingular symmetric finite form.* □

The homomorphism $Z \to Q/Z$ given by $k \longrightarrow k/p^n \mod Z$ embeds Z_{p^n} into $/\mathbb{Z}$. Thus the pairings φ^1,φ^2 can be realized as Z_{p^n}-valued by taking

$$\varphi^1([x],[y]) = \varphi(x,\theta y) \bmod p^n Z$$

$$\varphi^2([x],[y]) = \varphi(x,y) \bmod p^n Z .$$

If A is a free abelian C_{p^n}-module with a symmetric (skew symmetric) Z-valued inner product the finite form on $(H^i(C_{p^n};A),\varphi^i)$ can be considered as $\mathbb{Z}_{p^n}$-valued. An anisotropic Witt representative would have values in the $\mathbb{Z}_p \subset \mathbb{Z}_{p^n}$ and can therefore be thought of an an element of $W(F_p)$.

We want to compare φ^1 and φ^2 with the pairings defined in cohomology by (1.4)

$$H^1(C_{p^n};A) \times H^1(C_{p^n};B) \to H^2(C_{p^n};A \otimes B) \to H^2(C_{p^n};Z)$$

and

$$H^2(C_{p^n};A) \times H^2(C_{p^n};B) \to H^2(C_{p^n};A \otimes B) \to H^2(C_{p^n};Z)$$

where we intend to identify $H^2(C_{p^n};Z)$ with $Z/p^nZ = Z_{p^n}$. From [C-E], we find that if $[a] \in H^1(C_{p^n},A)$ and $[b] \in H^2(C_{p^n};B)$ then $[a] \cup [b]$ is represented by

$$\sum_{0\leq r<s\leq p^n} T^r a \otimes T^s b .$$

The coefficient homomorphism $\varphi_*: H^2(C_{p^n};A \otimes B) \to H^2(C_{p^n};Z)$ is given by $\varphi_*([\Sigma_i a_i \otimes b_i]) = \Sigma_i \varphi(a_i,b_i)$. The composition in (II.1.4) sends $([a],[b])$ to

$$\sum_{0\leq r<s\leq p^n} \varphi(T^r a,T^s b) = \varphi(a, \sum_{r<s} T^{s-r} b) \bmod p^n Z .$$

We leave it to the reader to show that

$$\sum_{r<s} T^{s-r} = T^{\frac{p^n+1}{2}} \theta + \frac{p^n+1}{2} \Sigma .$$

Therefore

$$\varphi(a, \sum_{r<s} b) = \varphi(a, T^{\frac{p^n+1}{2}} \theta b) + \varphi(a, \tfrac{p^n+1}{2} \Sigma b)$$

$$= \varphi^1([a], [T^{\frac{p^n+1}{2}} b])$$

but $T^{\frac{p^n+1}{2}} b \equiv b \bmod {}_{\Delta}B$ so that $[T^{\frac{p^n+1}{2}} b] = [b]$ and φ^1 agrees with the composition. Similarly for φ^2.

For any C_{p^n}-module A there are homomorphisms

$$\psi_1 : H^1(C_{p^n};A) \to H^2(C_{p^n};A/_{p^n}A)$$

and

$$\psi_1([a]) = [\theta a] \quad \text{and} \quad \psi_2([a]) = [\bar{a}]$$

where $\bar{a}$ is $a \bmod p^n A$. If A is p-torsion free then the sequence $0 \to A \xrightarrow{p} A \to A/_{p^n}A \to 0$ is exact and the six term cohomology sequence splits into two short exact sequences

$$0 \to H^1(C_{p^n};A) \to H^1(C_{p^n};A/_{p^n}A) \xrightarrow{\delta_2} H^2(C_{p^n};A) \to 0$$

$$0 \to H^2(C_{p^n};A) \to H^2(C_{p^n};A/_{p^n}A) \xrightarrow{\delta_1} H^1(C_{p^n};A) \to 0$$

and $\delta_2 \circ \psi_2 = 1$ and $\delta_1 \circ \psi_1 = 1$.

If $A \times B \to Z$ is a nonsingular, C_{p^n}-invariant pairing, there is an induced pairing

$$H^1(C_{p^n};A) \times H^2(C_{p^n};B/_{p^n}B) \to Z_{p^n}$$

given by $([a],[b]) \longrightarrow \varphi(a,b) \bmod p^n$. If $[b] \in H^1(C_{p^n};B)$ then $\psi_1[b] = [\theta b] \in H^2(C_{p^n},B/_{p^n}B)$. Then

$$\varphi(a,\theta b) \equiv \varphi(a,(T^{\frac{p^n+1}{2}}\theta + \frac{p^n+1}{2}\Sigma)b) \bmod p^n$$

$$\equiv \varphi(a,\sum_{r<s} t^{s-r}b) = \varphi_*[\sum_{r<s} t^r a \otimes t^s b]$$

$$\equiv \varphi^1([a],[b]) \bmod p^n .$$

Therefore $\varphi^1([a],[b]) = \varphi_*([a] \cup \psi_1[b])$. Similarly $\varphi^2([a],[b]) = \varphi_*([a] \cup \psi_2[b])$.

(1.5) Lemma: The maps

$$(C_{p^n}, A, \varphi) \longrightarrow (H^2(C_{p^n};A), \varphi^2)$$

$$(C_{p^n}, A, \varphi) \longrightarrow (H^1(C_{p^n};A), \varphi^1)$$

are well-defined homomorphisms of

$$W_o(Z, C_{p^n}) \longrightarrow W(F_p) \quad \text{and}$$

$$W_2(Z, C_{p^n}) \longrightarrow W(F_p)$$

respectively.

Proof: All we have to do is check that a metabolic module goes to a metabolic module. We do this in the second case. Suppose $U \subset A$, $U = U^\perp$, and U is invariant under the action of C_{p^n}. Let $\hat{U} = \{[u] \mid u \in U^\Sigma\} \subset H^1(C_{p^n};A)$. Clearly $\theta U \subset U$ so $\hat{U} \subset \hat{U}^\perp$. If $[x] \in H^1(C_{p^n};A)$ and $\varphi^1([x],[u]) = 0 \bmod Z$ for all $[u] \in \hat{U}$ then (since φ is nonsingular) there is a $y \in A$ such that $\varphi(y,u) = 1/p^n \cdot \varphi(x,\theta u)$ for all $u \in U^\Sigma$. As before this implies $\varphi(-\Delta y,\theta u) = \varphi(x,\theta u)$. Therefore $x + \Delta y$ annihilates $\theta(U^\Sigma)$, hence annihilates U^Σ. $x + \Delta y \in A^\Sigma$ so it annihilates $_\Sigma A$ and therefore $_\Sigma U$. An element that annihilates $_\Sigma U$ and U^Σ annihilates U. $x + y \in U^\perp = U$. This implies $[x] = [x + \Delta y] \in \hat{U}$ and $\hat{U} = \hat{U}^\perp$.

Notice that a metabolizer for $H^1(C_{p^n};A)$ is $\operatorname{Im}\{H^1(C_{p^n};U) \to H^1(C_{p^n};A)\}$ where

U is a metabolizer for A. A similar proof works for $H^2(C_{p^n};A)$. For actions of π where $T \in \text{Center}(\pi)$ the construction

$$(C_{p^n};A,\varphi) \longrightarrow (H^i(C_{p^n};A),\varphi^i)$$

is a well-defined homomorphism

$$W_0(Z,\pi) \longrightarrow W_0(F_p,\pi/C_{p^n}) , \quad i = 2$$

or $$W_2(Z,\pi) \longrightarrow W_0(F_p,\pi/C_{p^n}) , \quad i = 1 . \quad \square$$

These cohomology constructions are associated with boundaries of rational forms in the following way. Let T be a generator of C_{p^n} and define

$$\Sigma_s = \sum_{i=0}^{p^{n-s}-1} T^{ip^s} \quad \text{and} \quad \Delta_s = T^{\frac{p^{n-s}-1}{2}} - T^{\frac{p^{n-s}+1}{2}}$$

and θ_s the unique solution to the equation

$$\Delta_s\theta_s = p^{n-s} - \Sigma_s \quad \text{so that} \quad \hat{\theta}_s = -\theta_s .$$

Let $(C_{p^n},V,(,))$ be a symmetric C_{p^n}-invariant Z-valued inner product and set $V(s) = \ker \Sigma_s$ and $U(s) = \ker \Delta_s$. (Note: $V(s)$ and $U(s)$ are both Z summands of V.) There is a symmetric rational form defined on $U(s)$ by $(x,y)_s = \frac{1}{p^{n-s}}(x,y)$. Consider $\text{Im}\,\Sigma_s \subset U(s) \otimes Q$. If $x, y \in \text{Im}\,\Sigma_s$ then

$$\begin{aligned}(x,y)_s &= \frac{1}{p^{n-s}}(\Sigma_s x',\Sigma_s y') \\ &= \frac{1}{p^{n-s}}(\Sigma_s^2 x',y') \\ &= (\Sigma_s x',y') \in Z .\end{aligned}$$

So $\text{Im}\,\Sigma_s$ is an integer lattice in $U(s) \otimes Q$. If $y \in U(s)$ then $\Sigma_s y = p^{n-s}y$ so that $U(s) \subset (\mathrm{Im}\ \Sigma_s)^+$.

Any homomorphism of $\mathrm{Im}\ \Sigma_s$ into $p^{n-s}\ Z$ can be extended to a homomorphism of U_s into Z because $U(s)/\mathrm{Im}\ \Sigma_s = H^2(C_{p^{n-s}};V)$ has exponent p^{n-s} and can therefore be realized as an innerproduct with some $x \in V$. $\Delta_s x$ annihilates $\mathrm{Im}\ \Sigma_s$ and the kernel of the projection $V \to U(s)$, hence it is zero. Therefore every homomorphism of $\mathrm{Im}\ \Sigma_s$ into $p^{n-s}\ Z$ is realized by innerproducts with elements in $U(s)$. This means $U(s) = (\mathrm{Im}\ \Sigma_s)^+$. This proves

(1.6) Lemma: $\partial([U(s)\otimes Q]) = [H^2(C_{p^{n-s}};V)] \in W_o(F_p)$. □

Similarly if $(C_{p^n}, V, (,))$ is skew symmetric then the rational innerproduct on $V(s)$ would be symmetric. The finite form $H^1(C_{p^n}, V)$ is also symmetric. These forms are related by

(1.7) Lemma: $\partial([V(s)\otimes Q]) = [H^1(C_{p^{n-s}};V)] \in W_o(F_p)$.

Proof: The innerproduct on $V(s)$ is defined by

$$(x,y)_s = \frac{1}{p^{n-s}} \langle x, \theta_s y\rangle \ .$$

Now one shows that $(\mathrm{Im}\ \Delta_s)^+ = V(s)$. □

If T is an orientation preserving diffeomorphism of a compact, closed manifold M^{2m} of period p^n then $H^m(M;Z)/TOR$ is a C_{p^n}-module with an invariant symmetric form if m is even and an invariant skew symmetric form if m is odd. M/T^{p^s}, $0 \leq s \leq n$, is a rational Poincare duality space with an integral orientation defined by the projection $M \to M/T^{p^s}$. Denote by $\mathrm{sgn}(T^{p^s},M)$ the signature of the rational form on $H^m(M/T^{p^s};Q)$.

Let $V = H^m(M;Z)/TOR$ and define

$$w(T^{p^s},M) = \begin{cases} \langle V(s), (,)_s\rangle \in W(Q) & \text{if } m \text{ is odd} \\ \langle U(s), (,)_s\rangle \in W(Q) & \text{if } m \text{ is even}\ . \end{cases}$$

Notice that $w(T^{p^s},M)$ is not quite the rational Witt class of the inner product defined on M/T^{p^s}. For example when C_{p^n} acts trivially, $U(s) = V$ for all s and $w(T^{p^s},M) = \langle\frac{1}{p^{n-s}}\rangle\, w(M)$.

Also define

$$d(T^{p^s},M) = w(T^{p^s},M) - \operatorname{sgn}(T^{p^s},M)\cdot 1 .$$

Restating lemmas II.1.6 and II.1.7 for group actions

(1.8) Lemma: $\partial(d(T^{p^s},M)) = \langle H^m(C_{p^{n-s}}, H^m(M;Z)/TOR\rangle \in W_o(F_p)$. □

If π is a finite group and A is a left π-module then $\operatorname{Hom}(A,Z)$ and $\operatorname{Hom}(A,Q/Z)$ are both left π-modules with $(g\cdot\alpha)(a) = \alpha(g^{-1}\cdot a)$ where $\alpha \in \operatorname{Hom}(A,Z)$ or $\operatorname{Hom}(A,Q/Z)$, $g \in \pi$, and $a \in A$.

For future reference we make note of the following.

(1.9) Lemma: Let A^* denote $\operatorname{Hom}(A,Q/Z)$ or $\operatorname{Hom}(A,Z)$ if A is a torsion or a free abelian group respectively.

(a) If A is a finite C_{p^n}-module then $H^1(C_{p^n},A^*)$ is naturally isomorphic to $H^2(C_{p^n};A)^*$.

(b) If A is a finitely generated free abelian group then $H^2(C_{p^n};A^*)$ is naturally isomorphic to $H^2(C_{p^n};A)^*$ for $i = 1,2$.

Proof: As usual T is a generator of C_{p^n} and $\Delta = T^{\frac{p^n-1}{2}} - T^{\frac{p^n+1}{2}}$ and $\Sigma = 1 + T + \ldots + T^{p^n-1}$. Dualizing $A \xrightarrow{\Sigma} A \xrightarrow{\Delta} A$ gives $A^* \xrightarrow{\Delta^*} A^* \xrightarrow{\Sigma^*} A^*$. To prove (a), since dualizing is exact, we can make identifications $\operatorname{im}(\Delta^*) \approx (\operatorname{im}\Delta)^*$ and $\ker(\Sigma^*) \approx (\ker\Sigma)^*$. Also

$$0 \to (A/\ker\Delta)^* \to (A/\operatorname{im}\Sigma)^* \to (\ker\Delta/\operatorname{im}\Sigma)^* \to 0$$

is exact. So

$$\begin{aligned} H^2(C_{p^n};A)^* &= (\ker \Delta/\operatorname{im} \Sigma)^* \\ &= (A/\operatorname{im} \Sigma)^*/(A/\ker \Delta)^* \\ &= \ker(\Sigma^*)/\operatorname{im}(\Delta^*) \\ &= H^1(C_{p^n};A^*). \end{aligned}$$

(b) follows from lemma II.1.2 and the natural pairing of $A^* \times A$ into Z. □

2. Periodic maps

Suppose T is an orientation preserving diffeomorphism on a closed, compact, even dimensional manifold M^{2m} of period p^n. T^* makes $H^*(M;Z)/\mathrm{TOR}$ into a left C_{p^n}-module. For every $1 \leq k \leq n$ there is a nontrivial subgroup $C_{p^k} \subset C_{p^n}$ generated by $T^{p^{n-k}}$.

(2.1) Definition: $f(k,M) = \langle H^m(C_{p^k};H^m(M;Z)/\mathrm{TOR})\rangle \in W(F_p)$ and $f(0,M) = 0$.

$$q(T^{p^{n-k}},M) = pf(k,M) - f(k-1,M) \in W(F_p) .$$

To begin we prove

(2.2) Proposition: *$q(T,M)$ is a cobordism invariant vanishing on fixed point free actions of C_{p^n}.*

Proof: If (M,C_{p^n}) bounds (W,C_{p^n}) then $U = \operatorname{Im}(H^m(W;Z)/\mathrm{TOR} \to H^m(M;Z)/\mathrm{TOR})$ is a C_{p^n}-invariant, self-annihilating subspace of one-half the rank of $H^m(M;Z)/\mathrm{TOR}$. Therefore by (I.1.4), $\langle H^m(M;Z)/\mathrm{TOR}\rangle = 0 \in W_*(Z,C_{p^n})$. By (II.1.5), $f(n,M)$ and $f(n-1,M)$ are zero so $q(T,M) = 0$. In [C-F], Conner-Floyd show that a fixed point free action of C_{p^n} is cobordant to an extension of a $C_{p^{n-1}}$ action. This means if C_{p^n} acts fixed point freely on M, then (M,C_{p^n}) is cobordant to (M',C_{p^n}) where M' is the disjoint union of p copies of V and $C_{p^{n-1}}$ acts on V by $v \longrightarrow tv$. C_{p^n} acts on M' by $T(v_1,\ldots,v_p) = (v_2,\ldots,v_p,tv_1)$.

A simple calculation shows that

$$f(n,M) = \langle H^m(C_{p^{n-1}}; H^m(V;Z)/TOR)\rangle$$
$$f(n-1,M) = p\langle H^m(C_{p^{n-1}}; H^m(V;Z)/TOR)\rangle$$

so that $q(T,M) = pf(n,M) - f(n-1,M) = 0$. □

We will now show that $q(T,M)$ can be computed by restricting our attention to M where $H^*(M;Z)$ has no p-torsion. First recall that Ω_*^{SO} has a set of generators with no odd torsion [St]. Since $q(T,M)$ is a cobordism invariant vanishing on fixed point free actions it can be computed in terms of the fixed set F of C_{p^n} and the action of C_{p^n} in a normal tube about F. Call this information the local data. In [CF], Conner-Floyd showed that the cobordism ring of local data is a polynomial algebra over Ω_*^{SO}. The generators are described as follows. Let ζ_r be the normal bundle of $CP(r)$ in $CP(r+1)$. This is a complex line bundle. Pick $1 \leq \alpha < p^n$ and let $\lambda = \exp(2\pi i/p^n)$, then there is an action of C_{p^n} on ζ_r by multiplication by λ^α. The classes $[CP(r), \zeta_r, \alpha]$ are the polynomial generators.

(2.3) <u>Lemma</u>: <u>There is an integer</u> K <u>such that</u> $p^K(M, C_{p^n})$ <u>has local data in the same cobordism class as an action on a manifold with no odd torsion</u>.

Proof. Begin by considering the algebraic curve $R = \{[z_o, z_1, z_2] \in CP(2) \mid z_o^{p^n} + z_1^{p^n} + z_2^{p^n} = 0\}$. For $1 \leq \alpha < p^n$, C_{p^n} acts on R by $[z_o, z_1, z_2] \longrightarrow [z_o, z_1, \lambda^\alpha z_2]$. This action has p^n isolated fixed points. Denote R with this action by R_α. C_{p^n} also acts on $CP(r+1)$ by $[z_o, \ldots, z_{r+1}] \longrightarrow [\lambda^\alpha z_o, z_1, \ldots, a_{r+1}]$ with local data $(CP(0), \zeta_o, -\alpha)^r \cup (CP(r), \zeta_r, \alpha)$. Denote $CP(r+1)$ with this action by $(CP(r+1), \alpha)$. By taking connected sums about isolated fixed points

$$(R_\alpha)^{r+1} \cup (p^n)^{r+1}(CP(r+1), \alpha)$$

is a closed manifold with torsion free cohomology and an action of C_{p^n} whose local data is $p^{n(r+1)}(CP(r),\zeta_r,\alpha)$. Call this manifold $A(r,\alpha)$.

The local data for (M,C_{p^n}) is cobordant to a polynomial in $(CP(r),\zeta_r,\alpha)$ with coefficients in Ω_*^{SO}. A high power of p times a monomial in the $(CP(r),\zeta_r,\alpha)$ is realizable as the local data for an action on a product of $A(r,\alpha)$.

A high power of p times (M,C_{p^n}) has the same local data as a disjoint union of products of $A(r,\alpha)$ and elements of Ω_*^{SO}. Since there are generators of Ω_*^{SO} with no odd torsion this proves the lemma. □

(2.4) <u>Corollary</u>: $q(T,M)$ <u>can be calculated by considering</u> M <u>where</u> $H^*(M;Z)$ <u>has no odd torsion</u>.

Proof. From (II.2.3), $p^{2N}(M,C_{p^n})$ has local data cobordant to the local data in $M' = \bigcup_{i=1}^{s} V_i \times W_i$ where V_i is a product of $A(r,\alpha)$ and $W_i \in \Omega_*^{SO}$. We can assume possibly changing $p^{2N}(M,C_{p^n})$ by a cobordism that the local data is equal. Taking connected sums about the local data $X = p^{2N}(M,C_{p^n}) \# (-M',C_{p^n})$ is a closed manifold on which C_{p^n} acts fixed point freely. Therefore

$$p^{2N}\cdot M = M' + X$$

and

$$\begin{aligned} q(T,M) &= p^{2N}q(T,M) \quad \text{because } p^2 \equiv 1 \bmod 8 \text{ and } W(F_p) \text{ has exponent dividing } 4. \\ &= q(T',M') + q(T_X,X) \\ &= \sum_{i=1}^{s} q(T_i \times 1_i, V_i \times W_i) \quad \text{by II.2.2} \\ &= \sum_{i=1}^{s} q(T_i,V_i)\operatorname{sgn} W_i \quad \text{because the action on } W_i \text{ is trivial.} \end{aligned}$$

Hence to compute $q(T,M)$ it suffices to compute $q(T_i,V_i)$, but $H^*(V_i,Z)$ is torsion-free. □

<u>Remark</u>: If α is relatively prime to p then the action of C_{p^n} on $A(r,\alpha)$ is semi-free. The V_i are products of the $A(r,\alpha)$ so by grouping ones where α is relatively prime to p together we can further restrict our attention to product actions $(T' \times T'', V' \times V'')$ where C_{p^n} acts semi-freely on V' and $(T'')^{p^{n-1}} = 1_{V''}$

nd all fixed sets also have torsion free cohomology.

. The spectral sequence associated to the equivariant cohomology of M

M admits the structure of a simplicial complex so that C_{p^n} acts simplicially n such a way that an open simplex is either left fixed or sent to another disjoint pen simplex. The cellular chain complex of $M_{C_{p^n}} = M \times EC_{p^n}/T \times \tau$ is a bicomplex C-E]. Associated to any bicomplex is a spectral sequence that converges to the raded group corresponding to some filtration of $H^*(M_{C_{p^n}};Z)$ or $H^*(M_{C_{p^n}};Z)$. Ve describe in detail this spectral sequence and its relation to the E_1 spectral equence defined by Serre using cubical singular simplices. The two spectral equences are isomorphic at the E_1-term. This gives us the necessary identification f the product structure at the E_2-term to prove the important theorem of the ection -- II.3.4.

A nonnegative bicomplex is a collection $\{K_{k,\ell} | k,\ell \geq 0\}$ of abelian groups with wo families of homomorphisms $\partial: K_{k,\ell} \to K_{k-1,\ell}$ and $\partial': K_{k,\ell} \to K_{k,\ell-1}$ satisfying $^2 = 0$, $(\partial')^2 = 0$, and $\partial\partial' + \partial'\partial = 0$. The total complex of a bicomplex is $_k = \bigoplus_{i=0}^{k} K_{i,k-i}$ and $d: X_k \to X_{k-1}$ by $d(X_0,\ldots,X_k) = (\partial x_0 + \partial' x_1, \partial x_1 + \partial x_2, \ldots, \partial x_k)$. he relation on ∂, ∂' imply $d^2 = 0$. This total complex may be denoted TOT(K).

Define $Z^r_{k,\ell} = \{x \in K_{k,\ell} | \text{There exist } x_i \in X_{k-i,\ell+1} \text{ so that } x = y_o,$
$$\partial' x_o = 0, \partial x_o + \partial' x_1 = 0, \partial x_1 + \partial' x_2 = 0, \ldots, \partial x_{r-1} + \partial' x_r = 0\}$$

and $B^r_{k,\ell} = \{x \in K_{k,\ell} | \text{There exist } y_i \in K_{k-i,\ell+i+1} \text{ so that } \partial' y_r = 0,$
$$\partial y_r + \partial' y_{r-1} = 0, \ldots, \partial y_2 + \partial' y_1 = 0, \partial y_1 + \partial' y_o = x\} .$$

Notice that if $x \in B_{k,\ell}$, there by defining $x_o = x$, $x_1 = \partial y_o$, $x_2 = 0, \ldots, x_\ell = 0$ hen $x \in Z^{r'}_{k,\ell}$ for every r'. Now define

$$E^r_{k,\ell} = Z^{r-1}_{k,\ell}/B^{r-1}_{k,\ell}$$

nd $d^r_{k,\ell}: E^r_{k,\ell} \to E^r_{k-r,1+r-1}$ by $d^r([x]) = [\partial x_{r-1}]$ where $x = x_o$, $\partial' x_o = 0$, $x_o + \partial' x_1 = 0, \ldots, \partial x_{r-1} + \partial' x_r = 0$.

We leave the following to the reader

(3.1) Lemma: d^r is well-defined and $H_*(E^r,d^r) = E^{r+1}$. E^∞ is the bigraded module associated to the filtration of TOT(K) given by $F_q(\mathrm{TOT}(K)) = \oplus K_{k,\ell}$ where $k \leq q$. □

To compute the E^1-term, notice that $Z^o_{k,\ell} = \{x \in K_{k,\ell} | \partial' x = 0\}$ and $B^o_{k,\ell} = \{x \in K_{k,\ell} | x = \partial y, y \in K_{k,\ell+1}\}$. Hence $E^1 = H(K,\partial')$. Now $Z^1_{k,\ell} = \{x \in K_{k,\ell} | \partial' x = 0$ and $\partial x + \partial' x_1 = 0\}$. This means $[\partial x] = 0$ in $E^1_{k-1,\ell}$ if $x \in Z^1_{k,\ell}$ so that $E^2 = H(E^1,\partial) = H(H(K,\partial'),\partial)$. Thus the spectral sequence shows how closely the iterated homology approaches the actually homology of TOT(K).

The important example we want to consider is the bicomplex associated to $M_{C_{p^n}} = M \times EC_{p^n}/C_{p^n}$. The cell complex for EC_{p^n} has a $Z[C_{p^n}]$ in every nonnegative dimension and the boundary map is multiplication by Δ going from an odd to an even dimension and multiplication by Σ going from an even to an odd dimension. Since

$$C_*(M_{C_{p^n}}) = \{C_*(M) \otimes_Z C_*(EC_{p^n})\} \otimes_{Z[C_{p^n}]} Z ,$$

where the action inside the brackets is diagonal, we see that $C_*(M_{C_{p^n}})$ is the total complex of the bicomplex $K_{k,\ell} = C_\ell(M)$ and

$$\partial : K_{k,\ell} \to K_{k-1,\ell} \quad \text{by} \quad \partial(x) = \begin{cases} -\Delta x & k \text{ odd} \\ \Sigma x & k \text{ even} \end{cases}$$

and $\partial' : K_{k,\ell} \to K_{k,\ell-1}$ by $\partial'(x) = (-1)^k \partial_M(x)$ where ∂_M is the differential in $C_*(M)$. The spectral sequence gives

$$E^1_{k,\ell} = H_\ell(M)$$
$$E^2_{k,\ell} = H_k(C_{p^n}; H_\ell(M)) .$$

Similarly for the cohomology spectral sequence

$$E_1^{k,\ell} = H^\ell(M)$$

and

$$E_2^{k,\ell} = H^k(C_{p^n};H^\ell(M)) \ .$$

More critical is the product structure on the E_2-term of the cohomology spec-
ral sequence. From [S] the products on $C^*(M_{C_{p^n}})$ induce a multiplicative structure
n E_r for all r and $r=2$ the product defined on

$$H^k(C_{p^n};H^\ell(M)) \times H^{k'}(C_{p^n};H^{\ell'}(M))) \to H^{k+k'}(C_{p^n};H^{\ell+\ell'}(M))$$

s $(-1)^{k'\ell}$ times the composition

$$H^k(C_{p^n},H^\ell(M)) \times H^{k'}(C_{p^n};H^{\ell'}(M)) \to H^{k+k'}(C_{p^n};H^\ell(M) \otimes H^{\ell'}(M)$$
$$\to H^{k+k'}(C_{p^n};H^{\ell+\ell'}(M))$$

here the first homomorphism is given by cup products in cohomology and the last is
nduced by the cup product in M.

On the bicomplex $C^*(M_{C_{p^n}})$ define a map $\Theta: K^{k,\ell} \to K^{k-1,\ell}$ by

$$\Theta(x) = \begin{cases} \bar{x} & \text{if } k \text{ is even} \\ -\theta x & \text{if } k \text{ is odd} \end{cases}$$

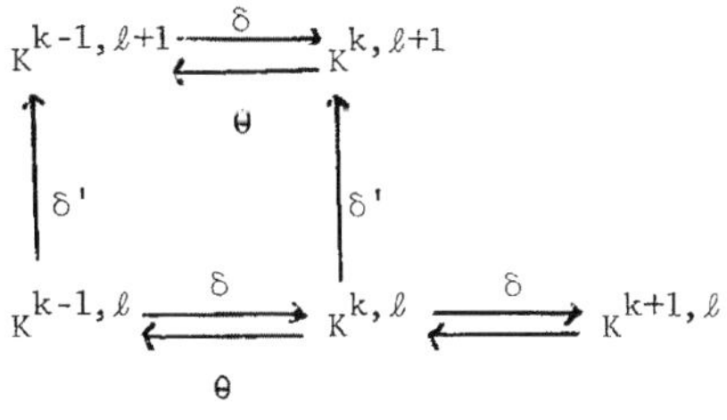

f k is even and $2m < k$, then

$$[\theta(\delta'+\delta) + (\delta'+\delta)\theta](x)$$
$$= \theta(\delta_M x - \Delta x) + (-\delta_M x + \Sigma x)$$
$$= \delta_M x + \theta\Delta x - \delta_M x + \Sigma x$$
$$= (\theta\Delta + \Sigma)x = p^n x .$$

Similarly if k is odd and $2m < k$, then

$$[(\theta(\delta'+\delta) + (\delta'+\delta)\theta]x = p^n x .$$

This map shows that $H_k(M_{C_{p^n}})$, $k > 2m$, is p-torsion of exponent p. Furthermore a direct computation shows that Θ induces a chain map of degree -1

$$\overline{\Theta}: C^*(M_{C_{p^n}}) \to C^{*-1}(M_{C_{p^n}}, M) \otimes Z_{p^n}$$

defined by $\overline{\Theta}(x) = (-1)^k \Theta(x)$, $x \in C^k(M_{C_{p^n}})$. $\overline{\Theta}$ preserves the filtrations and induces a map of spectral sequences

$$\overline{\Theta}_r: E_r^{k,\ell} \to \overline{E}_r^{k-1,\ell}$$

where $\overline{E}_r$ is the spectral sequence associated to the bicomplex whose total complex is

$$C^*(M_{C_{p^n}}, M) \otimes Z_{p^n} .$$

$\overline{\Theta}$ has the following important properties.

(3.2) Lemma:

a) *If β is the Z_{p^n} Bockstein then* $\beta\overline{\Theta}(x) = x$.

b) *The map induced on the E_2-term of the spectral sequence by $\overline{\Theta}$ is ψ_1 and ψ_2 defined in section* 1.

Proof:

a) Follows directly from the definitions of $\bar{\Theta}$ and β.

b) $\bar{\Theta}$ induces a map

$$\bar{\Theta}: E_2^{k,\ell} \to \bar{E}_2^{k-1,\ell}$$

where $E_2^{k,\ell} = H^k(C_{p^n}; H^\ell(M;Z))$ and $\bar{E}_2^{k-1,\ell} = H^{k-1}(C_{p^n}; H^\ell(M; Z_{p^n}))$. By the definition of Θ this map is ψ_1 or ψ_2 depending on whether k is odd or even. □

On E_r define the following pairing

$$\mu_r: E_r^{k,\ell} \times E_r^{k',\ell'} \to Z_{p^n}$$

$$\mu_r(a,b) = 0 \quad \text{if} \quad k+k' \neq 2s, \quad \ell+\ell' \neq 2m \ .$$

$$\mu_r(a,b) = \langle a \cup \bar{\Theta}_r b, [M \times S^{2s-1}/T \times \tau] \rangle$$

where $[M \times S^{2s-1}/T \times \tau]$ is the permanent cycle in $\bar{E}_r^{2s-1,2m}$ determined by the fundamental class of $M \times S^{2s-1}/T \times \tau$. Let $E_r^k = \bigoplus_{i=0}^{k} E_r^{i,k-i}$.

(3.3) Lemma: The finite group pairing

$$E_r^k \times E_r^{2m+2s-k} \to Z_{p^n}$$

is nonsingular for

$$2m + r - 1 \leq k \leq 2s - r \ .$$

Proof. First consider $r = 2$. Then, for $k \geq 2m+1$

$$E_2^k = \bigoplus_{i=0}^{k} H^i(C_{p^n}; H^{k-i}(M; Z))$$

and λ_2 is exactly the pairing defined in II.1.2. This starts our inductive argument. Now what we want to show is that $(B_r^{k+\ell})^\perp = Z_r^{k+\ell}$ for $2m+r \leq k+\ell \leq 2s-r+1$.

a) $Z_r^{k+\ell} \subset (B_r^{1+\ell})^\perp$.

Take $d_r x \in E_r^{k,\ell}$ where $x \in E_r^{k-r,\ell+r-1}$ and $y \in Z_r^{2s-k,2m-\ell}$, then $\mu_r(d_r x,y) = \langle d_r x \cup \overline{\Theta}_r y, [M\times S/T\times\tau]\rangle$. Since d_r is a derivation and y is a cycle, $d_r(x\cup\overline{\Theta}_r y) = d_r x\cup\overline{\Theta}_r y$ so that $\mu_r(d_r x,y) = \langle d_r(x\cup\overline{\Theta}_r y), [M\times S/T\times\tau]\rangle$, but $x\cup\overline{\Theta}_r y \in \overline{E}_r^{2s-r,2m+r-1}$. Since $r \geq 2$, $2m+r-1 > 2m$ so that $\overline{E}_r^{2s-r,2m+r-1} = 0$. Hence $\lambda_r(d_r x,y) = 0$ and $Z_r^{k+\ell} \subset (B_r^{k+\ell})^\perp$.

b) $(B_r^{k+\ell})^\perp \subset Z_r^{k+\ell}$.

Suppose $y \in (B_r^{k+\ell})^\perp$, that is $y \in E_r^{2s-k,2m-\ell}$ and $\mu_r(d_r x,y) = 0$ for all $x \in E^{k-r,\ell+r-1}$. Again using the fact that d_r is a derivation and $\overline{E}_r^{2s-r,2m+r-1} =$ $d_r x\cup\overline{\Theta}_r y = \pm\, x\cup\overline{\Theta}_r d_r y$, so that

$$\mu_r(d_r x,y) = \pm\,\mu_r(x\cup\overline{\Theta} d_r y) = 0$$

for all $x \in E_r^{k-r,\ell+r-1}$. This means $d_r y \in (E_r^{k-r,\ell+r-1})$ If the pairing

$$E_r^{k-r,\ell+r-1} \times E_r^{2s-k+r,2m-\ell-r+1} \to Z_{p^n}$$

is non-singular then $d_r y = 0$ and $y \in Z_r^{2s-k,2m-\ell}$. By induction this pairing is nonsingular if

$$2m + r-1 \leq k+\ell-1 \leq 2s-r ,$$

hence $\quad (B_r^{k+\ell})^\perp = Z_r^{k+\ell}$ if

$$2m+r \leq k+\ell \leq 2s-r+1 .$$

This means there is a well defined pairing $\hat{\mu}_r$ induced by μ_r on $E_{r+1}^{k+\ell} =$

$Z_r^{k+\ell}/B_r^{k+\ell}$. To finish the proof we have to show $\hat{\mu}_r = \mu_{r+1}$.

$$\begin{aligned}\hat{\mu}_r([x],[y]) &= \mu_r\langle x \cup \bar{\Theta}_r y, [M \times S/T \times \tau]\rangle \\ &= \langle [x] \cup [\bar{\Theta}_r y], [M \times S/T \times \tau]\rangle \\ &= \langle [x] \cup \bar{\Theta}_{r+1}[y], [M \times S/T \times \tau]\rangle \\ &= \mu_{r+1}([x],[y]) \ . \quad \square\end{aligned}$$

(3.4) Theorem: If $s \equiv m \bmod 2$ and $s \geq 4m$, then

$$f(n,M) = \langle Lk(M \times S/T \times \tau)] \in W(F_p)$$

where $H^*(M;Z)$ is torsion free.

Proof. Consider the pairings $\mu_2^{m+s}, \mu_3^{m+s}, \ldots, \mu_{2m+1}^{m+s} = \mu_\infty^{n+s}$ on E_r^{m+s}. These pairings are all non-singular as long as

$$2m + 2m + 1 \leq m + s \leq 2s - 2m$$

which is true if $s \geq 4m$. Now μ_2^{m+s} is the finite form defined on $\bigoplus_{i=0}^{2m} H^{m+s-i}(C_{p^n};H^i(M;Z))$ by II.1.2. If $U = \bigoplus_{i=0}^{m-1} H^{m+s-i}(C_{p^n};H^i(M;Z))$ then by Poincare duality on $H^*(M;Z)$ and the dimensions of elements, $U^\perp = U \oplus H^s(C_{p^n};H^m(M;Z))$.

Hence $U^\perp/U = H^s(C_{p^n};H^m(M;Z))$. This proves the Witt class of (E_2^{m+s}, μ_2^{m+s}) equals $f(n,M)$. By the proof in Lemma I.3.3 the Witt class of (E_r^{m+s}, μ_r^{m+s}) equals the Witt class of $(E_{r+1}^{m+s}, \mu_{r+1}^{m+s}) = ((B_r^{m+s})^\perp/B_r, \mu_r^{m+s})$. We are left with showing $\langle E_\infty^{m+s}, \mu_\infty^{m+s}\rangle = \langle Lk(M \times S/T \times \tau)\rangle$. E_∞^{m+s} is the bigraded module associated to a filtration of $H^{m+s}(M \times S/T \times \tau, Z)$. $E_\infty^{m+s,0}$ is an actual subgroup of $H^{m+s}(M \times S/T \times \tau, Z)$ and by dimensions $E_\infty^{m+s,0} \subset (E_\infty^{m+s,0})^\perp \subset H^{m+s}(M \times S/T \times \tau, Z)$. μ_∞^{m+s} induces a non-singular pairing

$$E_\infty^{m+s,0} \times E_\infty^{s-m,2m} \to Z_{p^n}$$

so that (again using dimensions)

$$H^{m+s}(M \times S/T \times \tau, Z) \,/\, (E_\infty^{m+s,0})^{\perp} \approx E_\infty^{s-m,2m} .$$

Now $E_\infty^{m+s-1,1}$ represents an actual subgroup in $(E_\infty^{m+s,0})^{\perp}/E_\infty^{m+s,0}$ that is self-annihilating. Proceeding inductively, we get that

$$\langle Lk(M \times S/T \times \tau)\rangle = \langle E_\infty^{s,m}, \mu_\infty^{s,m}\rangle$$

because $\mu_\infty^{s,m}$ is induced by the multiplicative structure of $C^*(M \times S/T \times \tau)$. Therefore

$$\begin{aligned}\langle Lk(M \times S/T \times \tau)\rangle &= \langle E_\infty^{s,m}, \mu_\infty^{s,m}\rangle \\ &= \langle E_\infty^{s+m}, \mu_\infty^{s+m}\rangle \\ &= f(n,M) . \quad \square\end{aligned}$$

Let us reconsider $X = M \times S^{2s-1}/T \times \tau$, for $s \geq 4m$.

We need more explicit information about $H^*(M \times S/T \times \tau; Z)$. If there is a fixed point $x_o \in M$ of the whole group C_{p^n}, then the bundle $M \times S/T \times \tau \xrightarrow{\rho} S/\tau = L$ has a section and $H^*(M \times S/T \times \tau; Z)$ contains $H^*(L; Z)$ as a summand. If $b \in H^2(L; Z)$ is the canonical generator then b^ℓ generates $H^{2\ell}(L; Z)$ for $1 \leq \ell \leq s-1$. Let $B = \rho^* b$.

(3.5) <u>Lemma</u>: $\cup B : H^j(M \times S/T \times \tau, \mathbb{Z}) \to H^{j+2}(M \times S/T \times \tau, \mathbb{Z})$ <u>is an isomorphism for</u> $4m \leq j \leq 2s - 2m - 2$.

Proof. Again look at the spectral sequence. At the E_2-term $\cup B$ induces an isomorphism

$$E_2^{q,r} \longrightarrow E_2^{q+2,r} \quad \text{for} \quad 0 < q < 2s - 2$$

that commutes with d_2, that is $d_2(x \cup B) = d_2(x) \cup B$. This means $\cup B : E_3^{q,r} \longrightarrow$

$E_3^{q+2,r}$ is an isomorphism for $1 < q < 2s-3$. Because $E_{2m+1} = E_\infty$, $\cup B\colon E_\infty^{q,r} \xrightarrow{\approx}$ $E_\infty^{q+2,r}$ for $2m\cdot 1 < q < 2s-2m-1$. Looking at the filtration

$$\cup B\colon H^j(M\times S/T\times\tau, Z) \longrightarrow H^{j+2}(M\times S/T\times\tau, Z)$$

is an isomorphism for $4m \leq j \leq 2s-2m-2$. $\square$

Suppose we pick i so that $4m \leq m+s-i \leq m+s+i \leq 2s-2m-2$. In this range the integral cohomology is all p-primary and

$$\cup B^i\colon H^{m+s-i}(M\times S/T\times\tau; Z) \to H^{m+s+i}(M\times S/T\times\tau; Z)$$

is an isomorphism. We can define a finite form on $H^{m+s-i}(M\times S/T\times\tau; Z)$ as follows. If β denotes the Z_{p^n} Bockstein then define

$$(x,y) = \langle x\cup\beta^{-1}(y)\cup B^i\,;\ [M\times S/T\times\tau]\rangle .$$

Denote the Witt class of this form by

$$\langle H^{m+s-i}(M\times S/T\times\tau; Z), B^i\rangle .$$

(3.6) <u>Lemma</u>: <u>For</u> s <u>and</u> i <u>large enough</u>

$$\begin{aligned}\langle H^{m+s-i}(M\times S/T\times\tau; Z), B^i\rangle &= \langle Lk(M\times S^{2s-2i-1}/T\times\tau)\rangle \\ &= \langle H^{s-i}(C_{p^n}; H^m(M; Z)/TOR)\rangle .\end{aligned}$$

<u>In particular, if</u> $s-i \equiv m \bmod 2$ <u>then</u>

$$\langle H^{m+s-i}(M\times S/T\times\tau; Z), B^i\rangle = f(n,M)$$

<u>and if</u> $s-i \not\equiv m \bmod 2$ <u>then</u>

$$\langle H^{m+s-i}(M\times S/T\times\tau;\mathbb{Z}),B^i\rangle = 0 .$$

Proof. The inclusion

$$i:\ M\times S^{2s-2i-1}/T\times\tau \longrightarrow M\times S^{2s-1}/T\times\tau$$

is an isomorphism in cohomology through dimensions $2s-2i-2$. The image of the fundamental class in $M\times S^{2s-2i-1}/T\times\tau$ is related to the fundamental class in $M\times S^{2s-1}/T\times\tau$ by $i_*([M\times S^{2s-2i-1}/T\times\tau]) = B^i\cap[M\times S^{2s-1}/T\times\tau]$. The lemma now follows from the definitions and II.3.4. □

4. Computation of $q(T,M)$ in terms of local data

This section is devoted to the computation of $q(T,M)$ in terms of local data. Before starting these results we must describe the orientations on various component of the fixed set F of C_{p^n} in M. Suppose C is a component of F. There is a complex structure on the normal bundle to C in M so that in each fibre the irreducible complex representations are given by multiplication by $\exp(2\pi i r_j/p^k)$ where $(r_j,p)=1$, $0<r_j\le\frac{p^k-1}{2}$, $1\le k\le n$. Call this the standard-orientation on the normal bundle and let $\alpha_C=\prod_j r_j$. Define an orientation on C as follows: If $p\equiv 3 \bmod 4$ then -1 is not a square mod p so we can change the complex structure so that each r_j is a square mod p. (If necessary change the complex structure so that r_j becomes $-r_j$.) This complex structure on the normal bundle along with the orientation on M induce an orientation for C. If $p\equiv 1 \bmod 4$, use the standard orientation on the normal bundle and the orientation on M to orient each component C of the fixed set F and then divide F into two parts F_1 and F_{-1} depending on whether the Legendre symbol $(\frac{\alpha_C}{p}) = \pm 1$. That is $C\subset F_1$ if and only if $(\frac{\alpha_C}{p}) = 1$.

(4.1) Theorem: With the fixed set oriented as described above

a) If $p\equiv 3 \bmod 4$, then $q(T,M) = p\,\mathrm{sgn}\,F\cdot\langle 1\rangle_p$.

b) If $p \equiv 1 \bmod 4$, then

$$q(T,M) = \operatorname{sgn} F_1 \cdot \langle 1\rangle_p + \operatorname{sgn} F_{-1} \cdot \langle\alpha\rangle_p$$

where F_i is the union of the components where $(\frac{\alpha_C}{p}) = i$, $\langle\gamma\rangle_p$ denotes the quadratic form γx^2 over F_p, and α is chosen so that $(\frac{\alpha}{p}) = -1$.

(4.2) Corollary: If $T^* = id$ on $H^m(M;Z)/TOR$ then $q(T,M) = p \operatorname{sgn} M \cdot \langle 1\rangle_p$ and therefore

a) $\operatorname{sgn} M \equiv \operatorname{sgn} F \bmod 4$ if $p \equiv 3 \bmod 4$

b) $\operatorname{sgn} M \equiv \operatorname{sgn} F_1 \bmod 2$

$\operatorname{sgn} F_{-1} \equiv 0 \bmod 2$ if $p \equiv 1 \bmod 4$.

If $0 \le k \le m$ then M/T^{p^k} is a rational Poincare duality space. For $m \equiv 0 \bmod 2$ define $\operatorname{sgn}(T^{p^k},M)$ to be the signature of the symmetric form defined on $H^m(M/T^{p^k};Q)$.

(4.3) Corollary:

$$p \operatorname{sgn}(T,M) - \operatorname{sgn}(T^p,M) = \begin{cases} (p-1)\operatorname{sgn} F \bmod 8 & \text{if } p \equiv 3 \bmod 4 \\ 4 \operatorname{sgn} F_1 \bmod 8 & \text{if } p \equiv 5 \bmod 8 \\ 4 \operatorname{sgn} F_{-1} \bmod 8 & \text{if } p \equiv 1 \bmod 8 . \end{cases}$$

For $n = 1$, II.4.1 and II.4.2 are proved in [AHV1] and [AHV2]. II.4.3 was proved in [C2] for $n = 1$.

By the results in section 2 to complete the proof we need to compute $q(T, V \times W)$ for a product action where T acts semifreely on V with fixed set F, $T^{p^{n-1}}\big|_W = 1_W$, $\operatorname{Fix}(T|_W) \ne \emptyset$, and F, V, and W all have torsion free cohomology.

(4.4) Lemma: For V, W, and F as described

$$q(T, V \times W) = \sum_{X \subset F} \operatorname{sgn} X \cdot \langle\alpha_x\rangle_p \cdot q(T_W, W) .$$

Proof. By II.3.4

$$f(n, V\times W) = Lk[V^{2v}\times W^{2w}\times S^{2s-1}/T\times\tau]$$

where $s \equiv v+w \bmod 2$. Since C_{p^n} acts freely on $(V-F)\times W\times S$ we see that

$$H^{v+w+s}(V\times W\times S/T\times\tau) \approx H^{v+w+s}(F\times W\times S/T\times\tau) .$$

(For simplicity we shall assume that F has only one component. Actually the calculations made here should be done for each component and summed up to obtain the total contribution of the local data.)

Let E be an equivariant tubular neighborhood around F in V. Choose the complex structure on the fibres as previously described. Consider the Thom-Gysin sequence of the sphere bundle

$$S^{2v-2f-1} \longrightarrow \dot{E}\times W\times S/T\times\tau \longrightarrow F^{2f}\times(W\times S/T\times\tau)$$

with Euler class χ.

$$(4.5)\quad \begin{array}{ccc} H^{v+w+s-1}(\dot{E}\times W\times S/T\times\tau) & \longrightarrow & H^{2f+w+s-v}(F\times(W\times X/T\times\tau)) \\ & & \downarrow \cup\chi \\ & & H^{v+w+s}(F\times(W\times S/T\times\tau)) \\ & & \downarrow \\ & & H^{v+w+s}(\dot{E}\times W\times S/T\times\tau) \ . \end{array}$$

$\dot{E}\times W\times S/T\times\tau$ is a sphere bundle in another way. Since the action on $\dot{E}$ is free $\dot{E}\times W/T$ is a manifold so that $S^{2s-1}\longrightarrow \dot{E}\times W\times S/T\times\tau \longrightarrow \dot{E}\times W/T$ is a sphere bundle. The Thom Isomorphism now implies that for $2v+uv < q < 2s-1$, $H^q(\dot{E}\times W\times S/T\times\tau;Z) = 0$. Therefore $\cup\chi$ in sequence II.4.5 is an isomorphism. As a submanifold of codimension 0, $(E,\dot{E})\times W \subset V\times W$, admits an orientation compatible with the orientation on $V\times W$. This orientation is passed on to an orientation on

$E, \dot{E}) \times W \times S/T \times \tau \subset V \times W \times S/T \times \tau$.

In the following diagram every map is an isomorphism.

(4.6)
$$\begin{array}{ccc}
H^{v+w+s}(V \times W \times S/T \times \tau;\, Z) & & \\
\downarrow{\scriptstyle i^*} & & \\
H^{v+w+s}(E \times W \times S/T \times \tau\ Z) & \xrightarrow{j^*} & H^{v+w+s}((E,\dot{E}) \times W \times S/T \times \tau\ Z) \\
\uparrow{\scriptstyle \rho^*} & & \uparrow{\scriptstyle \cup U} \\
H^{v+w+s}(F \times W \times S/T \times \tau\ Z) & \xrightarrow{\cup \chi} & H^{2f+w+s-v}(F \times W \times S/T \times \tau\ Z)
\end{array}$$

where U is the Thom class. Define a finite form on $H^{2f+w+s-v}(F \times (W \times S/T \times \tau); Z)$ by

$$\begin{aligned}
(a,b) &= \langle (a \cup U) \cup j^*(\beta^{-1}(b) \cup U),\ [(E,\dot{E}) \times W \times S/T \times \tau] \rangle \\
&= \langle (a \cup U) \cup \rho^*(\beta^{-1}(b) \cup \chi),\ [(E,\dot{E}) \times W \times S/T \times \tau] \rangle \\
&= \langle a \cup \beta^{-1}(b) \cup \chi,\ U \cap [(E,\dot{E}) \times W \times S/T \times \tau] \rangle \\
&= \langle a \cup \beta^{-1}(b) \cup \chi,\ F \times W \times S/T \times \tau \rangle .
\end{aligned}$$

Using the compatible orientations on $V \times W \times S/T \times \tau$ and $(E,\dot{E}) \times W \times S/T \times \tau$ and the isomorphism $(\cup U) \circ (j^*)^{-1} \circ i^*$ we see that the finite form just defined is isometric to $Lk(V \times W \times S/T \times \tau)$. Because $H^*(F;Z)$ is torsion free

$$H^{2f+w+s-v}(F \times (W \times S/T \times \tau)) = \bigoplus_{j=0}^{2f} H^j(F) \otimes H^{2f+w+s-v-j}(W \times S/T \times \tau).$$

If $U = \bigoplus_{j=f+1}^{2f} H^j(F) \otimes H^{2f+w+s-v-j}(W \times S/T \times \tau)$, then a simple check on the dimensions of elements shows $U \subset U^{\perp}$. In fact $U^{\perp} = \bigoplus_{j=f}^{2f} H^j(F) \otimes H^{2f+w+s-v-j}(W \times S/T \times \tau)$ and

(4.7) $$U^{\perp}/U \approx H^f(F) \otimes H^{w+s+f-v}(W \times S/T \times \tau) .$$

(4.8) Lemma: The Euler class χ has the form $\chi = 1 \otimes \alpha_F B^{v-f} + \sum_i f_i \otimes w_i$ where $\dim f_i > 0$.

Proof. Let $\chi = 1 \otimes \chi_0$ + other terms. To calculate χ_0 we want to calculate the Euler class of the pullback $W \times S/T \times \tau \longrightarrow F \times (W \times S/T \times \tau)$. But this is induced by the bundle

$$S^{2v-2f-1} \longrightarrow S^{2v-2f-1} \times S/T \times \tau \longrightarrow S/\tau = L$$

where T acts on $S^{2v-2f-1}$ by

$$T(z_1, \ldots, z_{v-f}) = (\lambda^{a_1} z_1, \ldots, \lambda^{a_{v-f}} z_{v-f}) \quad \text{and}$$

$\alpha_F = \Pi a_i$. This bundle is actually the sphere bundle of a sum of $v-f$ line bundles Each line bundle is the tensor product of the canonical line bundle over L with itself. The Euler class of this bundle can be computed by comparing it with the canonical line bundle over $CP(s-1)$. Since tensor products of line bundles add Euler classes and sums of line bundles multiply Euler classes we get $\chi_0 = \alpha_F B^{v-f}$. □

Using II.4.7, II.4.8, and the dimensions of elements we have reduced the finite form to the following form on $H^f(F) \otimes H^{w+s+f-v}(W \times S/T \times \tau)$.

$$\langle x \otimes y, x' \otimes y' \rangle = \langle x \cup x', [F] \rangle \langle y \cup \beta^{-1}(y') \cup \chi_0, [W \times S/T \times \tau] \rangle .$$

In $W_0(Q/Z)$ this is the Witt class of

$$\operatorname{sgn} F \cdot \langle H^{w+s+f-v}(W \times S/T \times \tau); \alpha_F B^{v-f} \rangle .$$

By II.3.6

$$\langle H^{w+s+f-v}(W \times S/T \times \tau); \alpha_F B^{v-f} \rangle = \begin{cases} \langle \alpha_F \rangle_p q(T|_W^W) & \text{if } f \text{ is even} \\ 0 & \text{if } f \text{ is odd} . \end{cases}$$

Since $\operatorname{sgn} F = 0$ if f is odd, we get

$$f(n,V\times W) = \operatorname{sgn} F\cdot\langle\alpha_F\rangle_p\cdot f(n,W) .$$

Similarly

$$f(n-1,V\times W) = \operatorname{sgn} F\cdot\langle\alpha_F\rangle_p f(n-1,W)$$

so that

$$\begin{aligned} q(T,V\times W) &= pf(n,V\times W) - f(n-1,V\times W)\\ &= \operatorname{sgn} F\cdot\langle\alpha_F\rangle_p\cdot\{pf(n,W) - f(n-1,W)\}\\ &= \operatorname{sgn} F\cdot\langle\alpha_F\rangle_p\; q(T|_W,W) . \quad\square \end{aligned}$$

We are left with calculating $q(T,W)$ where $T^{p^{n-1}} = 1_W$. Let $t \in C_{p^{n-1}}$ be a generator and suppose $C_{p^n} \longrightarrow C_{p^{n-1}}$ so that $T \longrightarrow t$.

(4.9) Lemma: $q(T,W) = q(t,W)$.

Proof. By II.1.8

$$f(n,W) = \partial(d(T,W)) \quad\text{and}\quad f(n-1,W) = \partial(d(T^p,W)) .$$

If $T^{p^{n-1}} = 1_W$ then $\omega(T,W) = \langle p\rangle \otimes \omega(t,W)$ and $\omega(T^p,W) = \langle p\rangle \otimes \omega(t^p,W)$. Therefore

$$\begin{aligned} q(T,W) &= \partial(pd(T,W) - d(T^p,W))\\ &= \partial(p\{\langle p\rangle \otimes \omega(t,W) - \operatorname{sgn}(t,W)\cdot 1\}\\ &\quad - \partial(\{\langle p\rangle \otimes \omega(t^p,W) - \operatorname{sgn}(t^p,W)\cdot 1\}\\ &= \partial(\langle p\rangle \otimes \{pd(t,W) - d(t^p,W)\})\\ &\quad + \partial(\{p \operatorname{sgn}(t,W) - \operatorname{sgn}(t^p,W)\}(\langle p\rangle-1)) . \end{aligned}$$

Now by induction assume that II.4.1 and II.4.3 have been proven for $C_{p^{n-1}}$. Then using the orientations described in the II.4.1: if $p \equiv 1 \bmod 4$ we get

$$q(T,W) = q(t,W) + 4 \text{ sgn } F_{\pm 1}\langle 1\rangle_p = q(t,W)$$

because if $V \in W_0(Z\,(1/p))$ and $\text{sgn } V = 0$, then $\partial(\langle p\rangle \otimes V) = \partial(V)$.

If $p \equiv 3 \bmod 4$ we get

$$\begin{aligned} q(T,W) &= -q(t,W) + 2 \text{ sgn} F\langle 1\rangle_p \\ &= -q(t,W) + 2q(t,W) \\ &= q(t,W) \end{aligned}$$

because if $V \in W_0(\ (1/p))$ and $\text{sgn } V = 0$, then $\partial(\langle p\rangle \otimes V) = -\partial(V)$. □

To finish the proof we merely have to calculate $q(T,M)$ where C_p acts trivially on M. In this case $f(1,M) = p \text{ sgn} M\cdot\langle 1\rangle_p$ and $f(0,M) = 0$ so that

$$q(T,M) = p \text{ sgn} M\cdot\langle 1\rangle_p \ .$$

Actually we can use II.4.1 to calculate $f(n,M) \in W(F_p)$. Since

$$q(T^{p^{n-s}},M) = pf(s,M) - f(s-1,M)$$

$1 \leq s \leq n$, with $f(0,M) = 0$ we find that

$$p^n f(n,M) = p^{n-1}q(T,M) + p^{n-2}q(T^p,M) + \ldots + q(T^{p^{n-1}},M) \ .$$

Since $p^2 \equiv 1 \pmod 8$ we can multiply through by p^n to obtain

$$f(n,M) = \sum_{s=1}^{n} p^{n+s-1} q(T^{p^{n-s}},M) \ .$$

If $p \equiv 3 \bmod (4)$ then $F(T^{p^{n-s}}) \subset M$ is the fixed point set of $T^{p^{n-s}}$ oriented as specified. For $p \equiv 1 (\bmod 4)$

$$F(T^{p^{n-s}}) = F_1(T^{p^{n-s}}) \cup F_{-1}(T^{p^{n-s}})$$

.s the specified decomposition of the fixed point set of $T^{p^{n-s}}$

(4.10) Corollary: If $p \equiv 3 \pmod 4$ then in $W(F_p)$

$$f(n,M) = \left(\sum_{s=1}^{n} (-1)^{n-s} \operatorname{sgn}(F(T^{p^{n-s}}))\langle 1\rangle_p\right.$$

while if $p \equiv 1 \pmod 4$

$$f(n,M) = \left(\sum_{s=1}^{n} \operatorname{sgn}(F_1(T^{p^{n-s}}))\right)\langle 1\rangle_p + \left(\sum_{s=1}^{n} \operatorname{sgn}(F_{-1}(T^{p^{n-s}}))\right)\langle \alpha\rangle_p . \quad \square$$

We substituted for $q(T^{p^{n-s}},M)$ according to II.4.1 and if $p \equiv 3 \pmod 4$ used $p^{n+s} \equiv (-1)^{n-s} \pmod 4$.

Now M/T can be regarded as a rational homology manifold with an orientation compatible with that of M. Thus if $m \equiv 0 \pmod 2$ we may speak of the signature of M/T. Our next objective will be the computation of $\operatorname{sgn} M/T \pmod 8$.

To do this consider first the short exact sequence

$$0 \longrightarrow Z \longrightarrow Z_{(2)} \longrightarrow Z_{(2)}/Z \longrightarrow 0$$

where $Z_{(2)}$ is the ring of integers localized (but not completed) at 2. By analogy with $W(Q/Z)$,

$$W(Z_{(2)}/Z) \simeq \sum_{\text{odd primes}} W(F_p)$$

and there is a short exact sequence

$$0 \longrightarrow W(Z) \longrightarrow W(Z_{(2)}) \xrightarrow{\partial} W(Z_{(2)}/Z) \longrightarrow 0 .$$

Now van der Blij's invariant [M-H] may be defined on $W(Z_{(2)})$. If W is an inner-product space over $Z_{(2)}$ then there is a Wu class $u \in W$ so that

$$(u,x) \equiv (x,x) \pmod 2$$

for all $x \in W$. While v is not unique the $b(W) = (u,u) \bmod Z_{(2)}/8\,Z_{(2)} \simeq Z/8Z$ is unique and depends only on the Witt class $\langle W\rangle \in W(Z_{(2)})$.

If we define $A\colon W(Z_{(2)}) \longrightarrow Z/8Z$ by $A(\langle W\rangle) = (\mathrm{sgn}(W) - b\langle W\rangle) \in Z/8Z$. It is well known that this difference will vanish on the image of $W(Z) \longrightarrow W(Z_{(2)})$, and so there results an induced homomorphism

$$\tilde{A}\colon W(Z_{(2)}/Z) \longrightarrow Z/8Z \ .$$

We are really concerned with the restriction $\tilde{A}\colon W(F_p) \longrightarrow Z/8Z$.

(4.11) Lemma: On $W(F_p)$ the homomorphism $\tilde{A}$ is given by

$$\tilde{A}(\langle 1\rangle_p) \equiv 1-p \pmod 8$$

If $p \equiv 5 \pmod 8$

$$\tilde{A}(\langle \alpha\rangle_p) \equiv 0 \pmod 8$$

and if $p \equiv 1 \pmod 8$

$$\tilde{A}(\langle \alpha\rangle_p) \equiv 4 \pmod 8 \ .$$

Proof. [AHV2] As usual $1 \leq \alpha < p$ is a quadratic non-residue mod p. The first assertion follows simply by noting

$$\partial(\langle p\rangle - \langle 1\rangle) = \langle 1\rangle_p \in W(F_p) \ .$$

The signature of $\langle p\rangle - \langle 1\rangle$ is 0 and van der Blij's invariant is $p-1 \pmod 8$. This completely determines $\tilde{A}\colon W(F_p) \longrightarrow Z/8Z$ if $p \equiv 3 \pmod 4$.

We consider now a reciprocity relation. If q, p are distinct primes, both odd of course, then

$$\tilde{A}(\langle q\rangle_p) + \tilde{A}(\langle p\rangle_q) = 1 - pq \bmod 8 .$$

This follows from

$$\partial(\langle pq\rangle - 1) = (\langle q\rangle_p, \langle p\rangle_q) \in W(F_p) \oplus W(F_q) .$$

Thus if we fix $p \equiv 1 \pmod 4$ and, by way of induction assume the formula is established for all primes less than p then we need only remark that there is an odd prime $q < p$ with $(\frac{q}{p}) = -1$. Then $(\frac{p}{q}) = -1$ also so $\tilde{A}(\langle p\rangle_q)$ is computed by the inductive assumption and this determines $\tilde{A}(\langle q\rangle_p)$. The reader may work out the various cases with $p \equiv 1$ or $5 \pmod 8$. □

Continuing with sgn M/T in $H^m(M;Z)/TOR$ we identified the kernel of Δ as U. We can always take the Wu class of $H^m(M;Z)/TOR$ to lie in U. First note that if $\sum x = 0$ then

$$0 = (x,\Sigma x) = (x,x) + \sum_{1}^{p^n - \frac{1}{2}} 2(x,T^j x)$$

so $(x,x) \equiv 0 \bmod 2$ if $\Sigma x = 0$. In general $p^{2n}(x,x) = (\Delta\theta x + \Sigma x, \Delta\theta x + \Sigma x) \equiv (\Sigma x, \Sigma x)$ mod 2. Since p is odd, $(x,x) \equiv (\Sigma x,\Sigma x)$. For any Wu class v,

$$(\Sigma u, x) = (u, \Sigma x) \equiv (\Sigma x, \Sigma x) \equiv (x,x)$$

and thus Σu is also a Wu class. We shall henceforth take $u \in U$.

Now on U we imposed the bilinear form $p^n(x,y)$ to obtain an element $\omega(T,M) \in W(Z(1/p))$. Plainly

$$b(\omega(T,M)) = p^n \operatorname{sgn} M \bmod 8 .$$

Also by I.1.8

$$\partial(\omega(T,M)) = f(n,M) .$$

Now $\operatorname{sgn}(T,M) = \operatorname{sgn} M/T = \operatorname{sgn}(\omega(T,M))$ so

$$\tilde{A}(f(n,M)) = \operatorname{sgn} M/T - p^n \operatorname{sgn} M \bmod 8.$$

If we combine II.4.10 with II.4.11 we have

(4.12) Corollary: If $p \equiv 3 \pmod 4$

$$\operatorname{sgn} M/T - p^n \operatorname{sgn} M \equiv \left(\sum_{s=1}^{n} (-1)^{n-s} \operatorname{sgn}(F(T^{p^{n-s}}))\right)(1-p) \bmod 8$$

while if $p \equiv 5 \pmod 8$

$$\operatorname{sgn} M/T - p^n \operatorname{sgn} M \equiv 4\left(\sum_{s=1}^{n} \operatorname{sgn} F_1(T^{p^{n-s}})\right) \bmod 8$$

and finally if $p \equiv 1 \bmod 8$

$$\operatorname{sgn} M/T - p^n \operatorname{sgn} M \equiv 4\left(\sum_{s=1}^{n} \operatorname{sgn} F_{-1}(T^{p^{n-s}})\right) \bmod 8 . \qquad \square$$

From this corollary II.4.3 follows.

Chapter III
Cyclotomic Number Fields

1. Hermitian forms over a cyclotomic field

In this section we review some of the algebraic number theory required for the eventual computation of $W_*(Z,C_n)$, where C_n is a cyclic group of odd order. Let $\lambda = \exp 2\pi i|n$ and let $Q(\lambda)$ be the nth cyclotomic number field. This admits complex conjugation as its involution. We shall concentrate on $\mathcal{H}_0(Q(\lambda))$ and $\mathcal{H}_2(Q(\lambda))$. Denote by $Q(\lambda+\lambda^{-1})$ the fixed field of conjugation. This is the maximal real sub ield. We use $Z(\lambda) \subset Q(\lambda)$ to denote the ring of algebraic integers in $Q(\lambda)$ and $Z(\lambda+\lambda^{-1}) \subset Q(\lambda+\lambda^{-1})$ is the ring of real integers. Remember $1, \lambda, \ldots, \lambda^{\phi(n)-1}$ is an integral basis for $Z(\lambda)$.

We use $\mathcal{P}$ for primes in $Z(\lambda)$ and $\mathfrak{p}$ for primes in $Z(\lambda+\lambda^{-1})$. For finite primes in $Z(\lambda+\lambda^{-1})$ there are three cases

1) inert: $\mathfrak{p}Z(\lambda) = \mathcal{P}$ is a prime
2) ramified: $\mathfrak{p}_o Z(\lambda) = \mathcal{P}_o^2$, $\mathcal{P}_o \subset Z(\lambda)$ a prime
3) $\mathfrak{p}Z(\lambda) = \mathcal{P}_1\mathcal{P}_2$ for distinct primes in $Z(\lambda)$.

The ramified case occurs only when $n = p^k$ a power of an odd prime. In the first two cases $\bar{\mathcal{P}} = \mathcal{P}$ and in the third $\mathcal{P}_2 = \bar{\mathcal{P}}_1$. We use $\mathfrak{p}_{\infty_1}, \ldots, \mathfrak{p}_{\infty_m}$, $m = \phi(n)/2$ to determine the infinite primes (all real) in $Q(\lambda+\lambda^{-1})$. Each extends to a complex infinite prime in $Q(\lambda)$. We shall frequently refer to the Archimedean orderings associated with these real infinite primes.

Let $\sigma = \lambda+\lambda^{-1} - 2 \in Q(\lambda+\lambda^{-1})$. Then $(\lambda^{(1-n)/2} - \lambda^{(n-1)/2})^2 = \sigma$. That is, $Q(\lambda)$ is obtained from $Q(\lambda+\lambda^{-1})$ by adjoining $\sqrt{\sigma}$. Put $\mu = \lambda^{(1-n)/2} - \lambda^{(n-1)/2}$, then $\bar{\mu} = \mu$ and $\mu Z(\lambda)$ is the different of $Q(\lambda)$ over $Q(\lambda+\lambda^{-1})$, while $\sigma Z(\lambda+\lambda^{-1})$ is the discriminant. If $n = p^k$ is an odd prime power then $\sigma Z(\lambda+\lambda^{-1})$ is the only prime ideal in $Z(\lambda+\lambda^{-1})$ that ramifies. If n is composite then μ is a unit and $Q(\lambda)$ is unramified over $Q(\lambda+\lambda^{-1})$ ([We]).

For $x \in Q(\lambda+\lambda^{-1})^*$ we shall consider the Hilbert symbols (x,σ) at all primes, finite and infinite, in $Q(\lambda+\lambda^{-1})$. If $\mathfrak{p}$ splits then $(x,\sigma)_{\mathfrak{p}} = 1$. For an infinite prime $(x,\sigma)_{\mathfrak{p}_{\infty_j}} = 1$ if and only if x is positive in the ordering associated to $\mathfrak{p}_{\infty_j}$. If $\mathfrak{p}$ is inert then $(x,\sigma)_{\mathfrak{p}} = 1$ if and only if $\mathrm{ord}_{\mathfrak{p}} x \equiv 0 \pmod 2$. Finally if $\mathfrak{p}_o$ ramified occurs we can write $x = v(-\sigma)^{\nu}$ with $u \in \mathcal{O}(\mathfrak{p}_o)^*$ a unit in the local ring. But $-\sigma = \mu\bar{\mu}$ so

$$(x,\sigma)_{\mathfrak{p}_o} = (v,\sigma)_{\mathfrak{p}_o}((-\sigma)^{\nu},\sigma)_{\mathfrak{p}_o} = (u,\sigma)_{\mathfrak{p}_o} = 1$$

if and only if u is a square in the residue field $\mathcal{O}(\mathfrak{p}_o)/(\sigma) \simeq F_p$.

A special case of the Hasse Cyclic Norm Theorem tells us x is a Hermitian square if and only if $(x,\sigma)_{\mathfrak{p}} = 1$ for all primes, finite and infinite, in $Q(\lambda+\lambda^{-1})$ ([O'M]). The realization theorem for Hilbert symbols tells us that we can realize by $(x,\sigma)_{\mathfrak{p}}$ any specification of the symbols which is 1 at almost all primes, is 1 at every prime that splits and for which 1 is the product of all assigned values ([O'M]).

Let us turn now to $\mathcal{H}_o(Q(\lambda))$ the Witt ring of Hermitian forms over $Q(\lambda)$. Thi is a ring using tensor product over $Q(\lambda)$. There is $rk: \mathcal{H}_o(Q(\lambda)) \to Z/2Z$ which to each Hermitian innerproduct assigns its rank over $Q(\lambda)$ mod 2. The kernel of rk is the fundamental ideal $J \subset \mathcal{H}_o(Q(\lambda))$. Let $\mathcal{N}$ be the quotient of $Q(\lambda+\lambda^{-1})^*$ by the subgroup of Hermitian squares. There is a discriminant

$$\text{dis}: J \longrightarrow \mathcal{N} \longrightarrow \{1\} .$$

If (V,h) is a Hermitian form of even rank $2n$ we choose a basis and represent h as a Hermitian matrix $[h]$ over $Q(\lambda)$. Then $\text{dis}(\langle V,h\rangle) = (-1)^n \det[h]$ in $\mathcal{N}$. This is a Witt class invariant, for if (V,h) is metabolic we can find a basis for V in which $[h] = \begin{pmatrix} 0 & I \\ I & 0 \end{pmatrix}$. The discriminant is 1 for such matrices. In fact

$$\text{dis}: J/J^2 \simeq \mathcal{N} .$$

Finally there is signature, or more properly, multisignature. This is a ring homomorphism

$$\text{Sgn}: \mathcal{H}_o(Q(\lambda)) \longrightarrow Z^{\phi(n)/2} .$$

A rank one form is given on $Q(\lambda)$ by $h^x(v,w) = x v \bar{w}$, $x \in Q(\lambda+\lambda^{-1})^*$. This defines $\langle x\rangle \in \mathcal{H}_o(Q(\lambda))$. Now Sgn assigns +1 to $\langle x\rangle$ for those orderings at which x is positive and -1 at orderings in which x is negative. This uniquely determines Sgn. On J^2 the multisignature maps isomorphically onto the subring of elements divisible by 4. Furthermore, on J

$$(\mathrm{dis}\langle V,h\rangle,\sigma)_{\mathfrak{p}_{\infty_j}} = (-1)^{\mathrm{Sgn}_j\langle V,h\rangle/2} .$$

All of these remarks about $\mathcal{H}_o(Q(\Lambda))$ are consequences of Landherr's study of Hermitian forms over algebraic number fields ([L]).

If $\mathcal{P} \subset Z(\Lambda)$ is a prime with $\overline{\mathcal{P}} = \mathcal{P}$ and if $\mathcal{O}(\mathcal{P}) \subset Q(\Lambda)$ is the local ring then $\mathcal{O}(\mathcal{P})$ is conjugation invariant and there is $\mathcal{H}_o(\mathcal{O}(\mathcal{P}))$. We have $0 \longrightarrow \mathcal{H}_o(\mathcal{O}(\mathcal{P})) \longrightarrow \mathcal{H}_o(Q(\Lambda))$. Furthermore, with respect to the $\mathcal{O}(\mathcal{P})$-involution induced on $Q(\Lambda)/\mathcal{O}(\mathcal{P})$ we can introduce $\mathcal{H}_o(Q(\Lambda)/\mathcal{O}(\mathcal{P}))$. That is, the Witt group of $Q(\Lambda)/\mathcal{O}(\mathcal{P})$-valued Hermitian innerproducts on finitely generated torsion $\mathcal{O}(\mathcal{P})$-modules. By appealing to anisotropics we find that

$$\mathcal{H}_o(Q(\Lambda)/\mathcal{O}(\mathcal{P}) \simeq \mathcal{H}_o(\mathcal{O}(\mathcal{P})/(\pi)) .$$

We want to have the short exact sequence

$$0 \longrightarrow \mathcal{H}_o(\mathcal{O}(\mathcal{P})) \longrightarrow \mathcal{H}_o(Q(\Lambda)) \xrightarrow{\partial_{\mathcal{P}}} \mathcal{H}_o(Q(\Lambda)/\mathcal{O}(\mathcal{P})) \longrightarrow 0 .$$

Let us first show

(1.1) Lemma: If $\mathcal{P} = \overline{\mathcal{P}}$ and $\mathcal{P} \cap Z(\Lambda + \lambda^{-1})$ is inert then $\mathcal{H}_o(Q(\Lambda)/\mathcal{O}(\mathcal{P})) \simeq Z/2Z$, but if $\mathcal{P}_o \cap Z(\Lambda + \lambda^{-1})$ is ramified then $\mathcal{H}_o(Q(\Lambda)/\mathcal{O}(\mathcal{P}_o)) = 0$.

Proof: If $\mathfrak{p} = \mathcal{P} \cap Z(\Lambda + \lambda^{-1})$ is inert then we can choose a local uniformizer $\pi = \overline{\pi} \in Q(\Lambda + \lambda^{-1})^*$ for $\mathcal{O}(\mathcal{P})$. An $\mathcal{O}(\mathcal{P})/(\pi)$ linear imbedding of $\mathcal{O}(\mathcal{P})/(\pi) \subset Q(\Lambda)/\mathcal{O}(\mathcal{P})$ is given by $d \longrightarrow \mathrm{cl}(d\pi^{-1})$. Since $\pi = \overline{\pi}$ this commutes with induced involutions and $\mathcal{H}_o(Q(\Lambda)/\mathcal{O}(\mathcal{P})) \simeq \mathcal{H}_o(\mathcal{O}(\mathcal{P})/(\pi)) \simeq Z/2Z$. If $n = p^k$ there is a different situation at the ramified prime. If $\mathcal{P}_o \cap Z(\Lambda + \lambda^{-1})$ is ramified then we may use μ as the local uniformizer of $\mathcal{O}(\mathcal{P}_o)$. Note $\overline{\mu} = -\mu$. Now $d \longrightarrow \mathrm{cl}(d\mu^{-1})$ will induce an $\mathcal{O}(\mathcal{P}_o)/(\mu)$ linear embedding of the residue field into $Q(\Lambda)/\mathcal{O}(\mathcal{P}_o)$, however the involution induced by conjugation on the residue field is the identity while the induced involution on its image in $Q(\Lambda)/\mathcal{O}(\mathcal{P}_o)$ is multiplication by -1. Hence $W_2(\mathcal{O}(\mathcal{P}_o)/(\mu)) \sim \mathcal{H}_o(Q(\Lambda)/\mathcal{O}(\mathcal{P}_o)) = 0$ because the residue field has odd characteristic. The homomorphism $\partial_{\mathscr{P}}: \mathscr{H}_0(Q(\Lambda)) \longrightarrow \mathscr{H}_0(Q(\Lambda)/\mathcal{O}(\mathscr{P}))$ is defined using $\mathcal{O}(\mathscr{P})$-lattices. In the inert case we take $x \in Q(\Lambda + \lambda^{-1})$ and write $x = u\pi^{\nu}$ with $u \in \mathcal{O}(\mathscr{P})^*$ and $\pi = \bar{\pi}$ a local uniformizer. This yields $\langle u\pi^{\nu}\rangle \in \mathscr{H}_0(Q(\Lambda))$. But $\langle u\pi^{\nu}\rangle = \langle u\rangle$ if $\nu \equiv 0 \pmod 2$ and $\langle u\pi^{\nu}\rangle = \langle u\pi\rangle$ if $\nu \equiv 1 \pmod 2$. Clearly $\partial_{\mathscr{P}}\langle x\rangle =$ iff $\mathrm{ord}_{\mathfrak{p}}(x) \equiv 0 \bmod 2$, iff $(x,\sigma)_{\mathfrak{p}} = 1$. The image of $\partial_{\mathscr{P}}\langle u\pi\rangle$ is the generator of $\mathscr{H}_0(Q(\Lambda)/\mathcal{O}(\mathscr{P})) \simeq Z/2Z$. In general a Hermitian innerproduct over $Q(\Lambda)$ can be diagonalized, and thus $\langle V,h\rangle = \langle x_1\rangle + \ldots + \langle x_k\rangle$ with $x_j \in Q(\Lambda + \lambda^{-1})^*$. Then $\partial_{\mathscr{P}}\langle V,h\rangle = 0$ iff $(x_1 x_2 \cdots x_k, \sigma)_{\mathfrak{p}} = 1$. This suggests a homomorphism

$$\det_{\mathfrak{p}} : \mathscr{H}_0(Q(\Lambda)) \longrightarrow Z^*$$

which can be defined at each inert prime $\mathfrak{p} \subset Z(\Lambda + \lambda^{-1})$. Represent (V,h) by a Hermitian matrix $[h]$ over $Q(\Lambda)$ and set

$$\det_{\mathfrak{p}} \langle V,h\rangle = (\det[h],\sigma)_{\mathfrak{p}} \in Z^* .$$

This is a Witt class invariant because -1 is a local norm at each inert prime. If $Z/2Z$ is identified with Z^* then

(1.2) <u>Lemma</u>: <u>If</u> $\mathscr{P} \cap Z(\Lambda + \lambda^{-1}) = \mathfrak{p}$ <u>is inert then</u> $\partial_{\mathscr{P}}: \mathscr{H}_0(Q(\Lambda)) \longrightarrow \mathscr{H}_0(Q(\Lambda))/\mathcal{O}(\mathscr{P}))$ <u>agrees with</u> $\det_{\mathfrak{p}} : \mathscr{H}_0(Q(\Lambda)) \longrightarrow Z^*$.

We might give another expalantion of what happens when $\mathscr{P}_0 \cap Z(\Lambda + \lambda^{-1})$ is ramified. We can use $-\sigma = \mu\bar{\mu}$ as the uniformizer of $\mathcal{O}(\mathfrak{p}_0) \subset Q(\Lambda + \lambda^{-1})$ and μ as the uniformizer in $\mathcal{O}(\mathscr{P}_0)$. But if $x = \bar{x} \in Q(\Lambda + \lambda^{-1})^*$ then $x = u(-\sigma)^{\nu}$ with $u \in \mathcal{O}(\mathfrak{p}_0)^*$. Since $-\sigma$ is a Hermitian square, however $\langle x\rangle = \langle u\rangle \in \mathscr{H}_0(Q(\Lambda))$. This means

$$\mathscr{H}_0(\mathcal{O}(\mathscr{P}_0)) \simeq \mathscr{H}_0(Q(\Lambda))$$

if $\mathscr{P}_0 \cap Z(\Lambda + \lambda^{-1})$ is ramified.

Naturally we want to pass over to

$$0 \longrightarrow \mathcal{H}_o(Z(\Lambda)) \longrightarrow \mathcal{H}_o(Q(\Lambda)) \xrightarrow{\partial} \mathcal{H}_o(Q(\Lambda)/Z(\Lambda)) .$$

The interesting point is that primes for which $\bar{\mathcal{P}} \neq \mathcal{P}$ make no contribution. This is seen as follows. Let (A,h) be a torsion Hermitian form with values in $Q(\Lambda)/Z(\Lambda)$ and assume (A,h) is anisotropic. For each $\mathcal{P} \subset Z(\Lambda)$ let

$$A(\mathcal{P}) = \{a \mid a \in A, \mathcal{P}a = \{0\}\} .$$

We claim $A(\mathcal{P}) = \{0\}$ if $\mathcal{P} \neq \bar{\mathcal{P}}$. Suppose $0 \neq a \in A(\mathcal{P})$ and $\alpha \in \mathcal{P}$ is an element for which $\bar{\alpha} \notin \mathcal{P}$. Then $\bar{\alpha}a \neq 0$ because $\mathcal{P}$ is maximal. On the other hand $h(b,\bar{\alpha}a) = h(\alpha b, a) = 0$ for all $b \in A(\mathcal{P})$. Thus $\bar{\alpha}a \in A(\mathcal{P}) \cap A(\mathcal{P})^{\perp}$ which contradicts the anisotropic assumption.

Therefore

$$\sum_{\bar{\mathcal{P}}=\mathcal{P}} \mathcal{H}_o(Z(\Lambda)/\mathcal{P}) \simeq \mathcal{H}_o(Q(\Lambda)/Z(\Lambda)) .$$

Furthermore, we may omit $\mathcal{P}_o$ if $\mathcal{P}_o \cap Z(\Lambda+\lambda^{-1})$ ramifies. Thus ∂ splits up into $\partial_{\mathcal{P}}$ with $\mathcal{P} = \bar{\mathcal{P}}$ over inert.

(1.3) Lemma: An element $\langle V,h\rangle \in \mathcal{H}_o(Q(\Lambda))$ lies in the image of $0 \longrightarrow \mathcal{H}_o(Z(\Lambda)) \longrightarrow \mathcal{H}_o(Q(\Lambda))$ if and only if for every inert prime $\mathfrak{p} \subset Z(\Lambda+\lambda^{-1})$

$$\det_{\mathfrak{p}}(\langle V,h\rangle) = 1 .$$

We can characterize the structure of $\mathcal{H}_o(Z(\Lambda))$. There is no torsion. A torsion element in $\mathcal{H}(Q(\Lambda))$ has $\mathrm{Sgn} = 0$ so its rank is even and it is determined uniquely by its discriminant in $\mathcal{N}$. So let $\tau \in J$ be a torsion class. If it is to lie in $\mathcal{H}_o(Z(\Lambda))$ then at every $\mathfrak{p} \subset Z(\Lambda+\lambda^{-1})$ which is inert $\det_{\mathfrak{p}}(\tau) = (\mathrm{dis}(\tau),\sigma)_{\mathfrak{p}} = 1$. From the vanishing of signatures, $(\mathrm{dis}\ \tau,\sigma)_{\mathfrak{p}_{\infty_j}} = 1$ at all infinite primes. If there is a $\mathfrak{p}_o$ which ramifies there is only one and $(\mathrm{dis}\ \tau,\sigma)_{\mathfrak{p}_o} = 1$ also by reciprocity. Hence $\mathrm{dis}\ \tau$ is $1 \in \mathcal{N}$ and $\tau = 0$.

Now suppose $\langle V,h\rangle \in J$ lies in the image of $0 \longrightarrow \mathcal{H}_o(Z(\Lambda)) \longrightarrow \mathcal{H}_o(Q(\Lambda))$. Then for each infinite prime $\mathfrak{P}_{\infty_j}$, $\mathrm{sgn}_j\langle V,h\rangle \equiv 0 \pmod 2$ and by Hilbert symbol reciprocity

$$\prod_1^{\phi(n)/2} (\mathrm{dis}\langle V,h\rangle,\sigma)_{\mathfrak{P}_{\infty_j}} = (-1)^{\sum (\mathrm{sgn}_j\langle V,h\rangle)/2} = \begin{cases} 1 & n \neq p^k \\ (\mathrm{dis}\langle V,h\rangle,\sigma)_{\mathfrak{P}_o} \end{cases}$$

where in the second case $n = p^k$ and $\mathfrak{P}_o$ is the prime that ramifies.

(1.4) <u>Theorem</u>: *The Witt group* $\mathcal{H}_o(Z(\Lambda))$ *is torsion free. If* $n = p^k$ *then the image of*

$$\mathrm{Sgn}\colon \mathcal{H}_o(Z(\Lambda)) \longrightarrow Z^{\phi(n)/2}$$

is those integral $\phi(n)/2$*-tuples* $(m_1,\ldots,m_{\phi(n)/2})$ *with* $m_1 \equiv \ldots \equiv m_{\phi(n)/2} \pmod 2$. *If* n *is composite we must add*

$$m_1 + \ldots + m_{\phi(n)/2} \equiv 0 \pmod 4 \quad \text{if} \quad m_1 \equiv 0 \pmod 2$$
$$m_1 + \ldots + m_{\phi(n)/2} \equiv \phi(n)/2 \pmod 4 \quad \text{if} \quad m_1 \equiv 1 \pmod 2 .$$

Proof: In the first case we can satisfy Hilbert reciprocity and choose $x_1,\ldots,x_{\phi(n)/2}$ in $Q(\Lambda+\lambda^{-1})^*$ so that $\langle x_j,\sigma\rangle = 1$ for all inert $\mathfrak{P}$ and $x_j < 0$ at $\mathfrak{P}_{\infty_j}$ while $x_j > 0$ if $i \neq j$. Each $\langle x_j\rangle$ lies in the image of $\mathcal{H}_o(Z(\Lambda)) \longrightarrow \mathcal{H}_o(Q(\Lambda))$. The subring generated by all these $\mathrm{Sgn}(\langle x_j\rangle)$ is the subring described as the image of $\mathrm{Sgn}\colon \mathcal{H}_o(Z(\Lambda)) \longrightarrow Z^{\phi(n)/2}$.

In the second case we can do the following. If $1 \leq i < j \leq \phi(n)/2$ there is $X(i,j)$ in $Q(\Lambda+\lambda^{-1})^*$ with $(X(i,j),\sigma) = 1$ at every inert $\mathfrak{P}$, $X(i,j)$ is negative at $\mathfrak{P}_{\infty_i}$, $\mathfrak{P}_{\infty_j}$ and positive elsewhere. Each $\langle X(i,j)\rangle$ is in the image of $\mathcal{H}_o(Z(\Lambda)) \longrightarrow \mathcal{H}_o(Q(\Lambda))$ and the subring generated by the $\mathrm{Sgn}(X(i,j))$ is as ndicated. □

Now we are also concerned with an over-ring of $Z(\Lambda)$. Let $N > 1$ be a fixed nteger. A finite family $\mathcal{F}(N)$ of prime ideals is described by $\mathcal{P} \in \mathcal{F}(N)$ if and nly if $\mathcal{P} | NZ(\Lambda)$. Let

$$D(N) = \bigcap_{\mathcal{P} \notin \mathcal{F}(N)} \mathcal{O}(\mathcal{P}) .$$

learly $Z(\Lambda) \subset D(N) \subset Q(\Lambda)$, $D(N)$ is a Dedekind domain and $D(N)$ is stable under onjugation. Thus we have $0 \longrightarrow \mathcal{H}(D(N)) \longrightarrow \mathcal{H}_o(Q(\Lambda)) \xrightarrow{\partial} \mathcal{H}_o(Q(\Lambda)/D(N))$. Actually o new analysis is required since $\mathcal{H}_o(Q(\Lambda)/D(N)) \simeq \sum_{\bar{\mathcal{P}}=\mathcal{P}\notin\mathcal{F}} \mathcal{H}_o(Q(\Lambda)/\mathcal{O}(\mathcal{P}))$. The prime deals in $D(N)$ are $D(N) \cap \mathcal{O}(\mathcal{P})$ where $\mathcal{P} \notin \mathcal{F}(N)$. Thus

(1.5) *Lemma: If $\langle V,h\rangle \in \mathcal{H}_o(Q(\Lambda)$ then $\langle V,h\rangle$ lies in the image of*

$$0 \longrightarrow \mathcal{H}_o(D(N)) \longrightarrow \mathcal{H}_o(Q(\Lambda))$$

f and only if $\det_{\mathfrak{p}} \langle V,h\rangle = 1$ at every $\mathfrak{p}$ inert for which $\mathcal{P} = \mathfrak{p} Z(\Lambda)$ does not elong to $\mathcal{F}(N)$.

We are therefore allowed to specify Hilbert symbols at infinity and at inert rimes $\mathfrak{p}$ with $\mathcal{P} \in \mathcal{F}(N)$ in any manner we please. Of course $\mathcal{H}_o(D(N))$ may well ave torsion. In fact

(1.6) *Theorem: If $n(N)$ is the number of inert for which $\mathcal{P} \in \mathcal{F}(N)$ then the torsion subgroup of $\mathcal{H}_o(D(N))$ is an elementary abelian 2-group hose 2-rank is $n(N)-1$. If $n(N) > 0$ then the image of $\mathrm{Sgn}: \mathcal{H}_o(D(N)) \longrightarrow Z^{\phi(n)/2}$ onsists of $(m_1,\ldots,m_{\phi(n)/2})$ with $m_1 \equiv \ldots \equiv m_{\phi(n)/2} \pmod 2$.*

The proof is left to the reader. It may be noted that if $p^k = n$ then $\mathcal{H}_o(D(p))$ remains torsion free and in fact $\mathcal{H}_o(Z(\Lambda)) \simeq \mathcal{H}_o(D(p))$.

It is appropriate now to discuss the skew case. We have $\mathcal{H}_*(Q(\Lambda)) = \mathcal{H}_o(Q(\Lambda)) \oplus \mathcal{H}_2(Q(\Lambda))$. At the end of section 1 we briefly introduced products. The idea here is the same using tensor products over $Q(\Lambda)$. For two skew classes we still insist on

$\langle V\otimes V_1, -(h\otimes h_1)\rangle \in \mathcal{H}_o(Q(\Lambda))$. Actually there is a good reason for this. We have $\langle\mu\rangle \in \mathcal{H}_2(Q(\Lambda))$ which is the skew-Hermitian form on $Q(\Lambda)$ given by $(v,w) \longrightarrow \mu v\bar{w}$. According to our definition

$$\langle\mu\rangle^2 = \langle -\mu^2\rangle = \langle\mu\bar{\mu}\rangle = 1 \in \mathcal{H}_o(Q(\Lambda)) .$$

That is, $\langle\mu\rangle$ is a unit. Furthermore, if $\mathcal{H}_2(Q(\Lambda)) \xrightarrow[\simeq]{\otimes\langle\mu\rangle} \mathcal{H}_o(Q(\Lambda))$ is used to extend Sgn to $\mathcal{H}_2(Q(\Lambda))$ we do get a ring homomorphism

$$\mathrm{Sgn}: \mathcal{H}_*(Q(\Lambda)) \longrightarrow Z^{\phi(n)/2}(C_2) .$$

That is $(\langle V,h\rangle, \langle V_1,h_1\rangle) \longrightarrow \mathrm{Sgn}\langle V,h\rangle I + \mathrm{Sgn}\langle V,-\mu h_1\rangle T$. If $n \neq p^k$ then μ is a skew unit so that $\mathcal{H}_2(Z(\Lambda)) \xrightarrow[\simeq]{\otimes\mu} \mathcal{H}_o(Z(\Lambda))$. If $n = p^k$ then $\mathcal{H}_2(D(p)) \xrightarrow[\simeq]{\otimes\mu} \mathcal{H}_o(D(p))$ since μ becomes a skew-unit in $D(p)$.

Actually about the only other comment we need is when $n = p^k$ and $\mathcal{P}_o \cap Z(\Lambda+\lambda^{-1}) = \mathfrak{p}_o$ ramifies. In that case

$$0 \to \mathcal{H}_2(\mathcal{O}(\mathcal{P}_o)) \longrightarrow \mathcal{H}_2(Q(\Lambda)) \xrightarrow{\partial\mathcal{P}_o} \mathcal{H}_2(Q(\Lambda)/\mathcal{O}(\mathcal{P}_o)) .$$

As we should expect from the discussion following III.1.2 in this case $W(\mathcal{O}(\mathcal{P}_o)/(\mu)) \simeq \mathcal{H}_2(Q(\Lambda)/\mathcal{O}(\mathcal{P}_o))$.

Let us see how this works. For $x \in Q(\Lambda+\lambda^{-1})^*$ there is $\langle\mu x\rangle \in \mathcal{H}_2(Q(\Lambda))$. Now $x = u(-\sigma)^\nu$ where $u \in \mathcal{O}(\mathfrak{p}_o)^*$. Since $(-\sigma)^\nu$ is a Hermitian square we may as well look at $u \in \mathcal{O}(\mathfrak{p}_o)^*$. Then $\mathcal{O}(\mathcal{P}_o) \subset Q(\Lambda)$ is an $\mathcal{O}(\mathcal{P}_o)$ lattice for the form $(v,w) \longrightarrow \mu u v\bar{w}$. Plainly the dual lattice is $\mu^{-1}\mathcal{O}(\mathcal{P}_o)$. But $\mu^{-1}\mathcal{O}(\mathcal{P}_o)/\mathcal{O}(\mathcal{P}_o)$ is identified with the residue field $\mathcal{O}(\mathcal{P}_o)/(\mu)$ via $d \longrightarrow \mathrm{cl}(d\mu^{-1})$. If $d_1, d_2 \in \mathcal{O}(\mathcal{P}_o)$ the $(\mu^{-1}d_1, \mu^{-1}d_2) \longrightarrow u\mu\mu^{-1}d_1\bar{\mu}^{-1}d_2 = -u\mu d_1 d_2$. Thus on $\mathcal{O}(\mathcal{P}_o)/(\mu)$ we receive $\partial_{\mathcal{P}_o}\langle\mu u\rangle = \langle -u\rangle \in W(\mathcal{O}(\mathcal{P}_o)/(\mu))$. In particular $\partial_{\mathcal{P}_o}$ will preserve rank mod 2 in the skew case

(1.7) <u>Lemma</u>: <u>The</u> $\partial_{\mathcal{P}_o}: \mathcal{H}_2(Q(\Lambda))$ $W(\mathcal{O}(\mathcal{P}_o)/(\mu))$ <u>preserves rank mod</u> 2. <u>On</u>

$\langle\mu\rangle\cdot J \subset \mathcal{H}_2(Q(\Lambda))$, $\partial_{\mathcal{P}_0}$ agrees with $J \longrightarrow Z^*$ given by

$\langle V,h\rangle \longrightarrow (\mathrm{dis}\langle V,h\rangle,\sigma)_{\mathfrak{p}_0} \in Z^*$.

Proof: We identify Z^* with the fundamental ideal in $W(\mathcal{O}(\mathcal{P}_0)/(\mu))$ where an element is determined by its discriminant mod squares. Suppose an element in J is diagonalized as $\langle x_1\rangle + \ldots + \langle x_{2m}\rangle$ with $x_j \in Q(\Lambda+\lambda^{-1})^*$. Now $x_j = u_j(-\sigma)^{\nu_j}$ and $(x_j,\sigma)_{\mathfrak{p}_0} = (u_j,\sigma)_{\mathfrak{p}_0}$ so we may as well take $\partial_{\mathcal{P}_0}(\langle\mu u_1\rangle + \ldots + \langle\mu u_{2m}\rangle) = \langle -u_1\rangle + \ldots + \langle -u_{2m}\rangle$ in $W(\mathcal{O}(\mathcal{P})/(\mu))$. This discriminant is $(-1)^m u_1 \ldots u_{2m}$. So this is 1 if and only if $(-1)^m u_1 \ldots u_m$ is a square mod (μ). Which is if and only if $((-1)^m x_1 \ldots x_{2m},\sigma)_{\mathfrak{p}_0} = 1 = (\mathrm{dis}\langle V,h\rangle,\sigma)_{\mathfrak{p}_0}$.

This says that the image of $\mathcal{H}_2(Z(\Lambda)) \longrightarrow \mathcal{H}_2(Q(\Lambda))$ lies in $\langle\mu\rangle J$. That is, if $n = p^k$ there are no rank 1 elements in $\mathcal{H}_2(Z(\Lambda))$. Actually we can define therefore discriminant on all $\mathcal{H}_2(Z(\Lambda))$. Using our rule for products, we assign to $\langle V,h\rangle \in \mathcal{H}_2(Z(\Lambda))$

$$\mathrm{dis}(\langle\mu\rangle\langle V,h\rangle) = \mathrm{dis}(\langle V,-\mu h\rangle) \in \mathcal{N} .$$

The reader may be interested in working out the image of $\mathrm{Sgn}\colon \mathcal{H}_*(Z(\Lambda)) \longrightarrow \oplus^{\phi(n)/2}(C_2)$ when $n = p^k$. We do get a new restriction on the image of $\mathcal{H}_2(Z(\Lambda))$.

. Hermitian forms over cyclotomic integers

We have discussed the image of $0 \longrightarrow \mathcal{H}_0(Z(\Lambda)) \longrightarrow \mathcal{H}_0(Q(\Lambda))$. In this section we want to take up the construction of $Z(\Lambda)$-valued Hermitian and skew-Hermitian, forms on projective $Z(\Lambda)$-modules.

Let us begin with the fractional ideals in $Q(\Lambda)$. If $A \subset Q(\Lambda)$ is a fractional ideal let

$$\hat{A} = \bar{A}^{-1} = \overline{(A^{-1})} .$$

Then $\hat{A}$ is again a fractional ideal. The meaning is this. If to $\mathrm{Hom}_{Z(\Lambda)}(A, Z(\Lambda))$ we give the $Z(\Lambda)$-module structure $(t\psi)(a) = \psi(\bar{t}a)$ for $t \in Z(\Lambda)$, $\psi\colon A \longrightarrow Z(\Lambda)$

and $a \in A$, then $\bar{A}^{-1}$ is isomorphic as a $Z(\lambda)$-module to $\mathrm{Hom}_{Z(\lambda)}(A, Z(\lambda))$. For $b \in \bar{A}^{-1}$ we let $\psi_b(a) = a\bar{b} \in Z(\lambda)$. Of course $A \longrightarrow \hat{A}$ is an involution of the fractional ideal group which extends $x \longrightarrow \bar{x}^{-1}$ on $Q(\lambda)^*$. Thus we find A is $Z(\lambda)$ isomorphic to $\mathrm{Hom}_{Z(\lambda)}(A, Z(\lambda))$ if and only if there is an $x \in Q(\lambda)^*$ such that $A = x\hat{A}$; that is, if and only if A and $\hat{A}$ belong to the same ideal class. Equivalently, if and only if $A\bar{A} = xZ(\lambda)$. We shall prove a key lemma.

(2.1) Lemma: If A and $\hat{A}$ belong to the same ideal class then there is a $y \in Q(\lambda + \lambda^{-1})^*$ such that

$$A\bar{A} = yZ(\lambda) .$$

Proof: We have an $x \in Q(\lambda)^*$ with $A\bar{A} = xZ(\lambda)$. But then $xZ(\lambda) = \bar{x}Z(\lambda)$ so that $\bar{x}x^{-1} = u \in Z(\lambda)^*$. For this unit, $u\bar{u} = 1$ and hence u is a root of unity. Thus $u = \pm \lambda^j$ for some $0 \leq j < n$. Because n is odd we can always find i, $0 \leq i < n$ with $2i \equiv j \bmod n$. Put $y = x\lambda^{-i}$. Then $\bar{y}y^{-1} = \bar{x}\lambda^i x^{-1}\lambda^i = \pm 1$. If $\bar{y}y^{-1} = +1$ we are done. If $\bar{y}y^{-1} = -1$ and n is composite replace y by $y\mu$ and we are done. In case $n = p^k$ we must show $\bar{y}y^{-1} = -1$ is impossible.

The trick is to localize at the ramified prime. Remember μ is the uniformizer. Further $A \cdot \mathcal{O}(\mathcal{P}_0) = (\mu^k) \subset Q(\mathcal{P}_0)$. Now $(\mu^k \cdot (-\mu)^k)\mathcal{O}(\mathcal{P}_0) = y\mathcal{O}(\mathcal{P}_0)$ so $yu = (-\sigma)^k$ with $u \in \mathcal{O}(\mathcal{P}_0)^*$. But since $\bar{y} = -y$ it follows $-u = \bar{u}$. This is impossible because $\mathcal{O}(\mathcal{P}_0)/(\mu)$, on which conjugation induces the identity, has odd characteristic. □

(2.2) Lemma: If $A\bar{A} = yZ(\lambda)$ with $\bar{y} = y$ then

$$[a', a] = a'\bar{a}/y \in Z(\lambda)$$

is a Hermitian innerproduct on the fractional ideal A.

Proof: If $\psi: A \longrightarrow Z(\lambda)$ is a $Z(\lambda)$-module homomorphism then there is $b \in \bar{A}^{-1}$ with $\psi(a') = a'\bar{b}$. But $A = y\bar{A}^{-1}$ so there is $a \in A$ with $a = yb$, or $\bar{b} = ay^{-1}$. □

Thus every rank 1, $Z(\Lambda)$ Hermitian innerproduct is given by a pair (y,A) with $A\bar{A} = yZ(\Lambda)$ and $y = \bar{y}$ in the manner prescribed. In other words $\langle A, [\ ,\ ]\rangle \in \mathcal{H}_0(Z(\Lambda))$ and $\langle y\rangle \in \mathcal{H}_0(Q(\Lambda))$ is its image.

Then given $y = \bar{y} \in Q(\Lambda + \lambda^{-1})^*$ we ask if there is a fractional ideal $A \subset Q(\Lambda)$ with $A\bar{A} = yZ(\Lambda)$.

(2.3) <u>Lemma</u>: <u>If</u> $y \in Q(\Lambda + \lambda^{-1})^*$ <u>then there is a fractional ideal</u> $A \subset Q(\Lambda)$ <u>with</u> $A\bar{A} = yZ(\Lambda)$ <u>if and only if at every inert</u> $\mathfrak{p} \subset Z(\Lambda + \lambda^{-1})$

$$(y,\sigma)_{\mathfrak{p}} = 1 .$$

Proof. Suppose first $(y,\sigma)_{\mathfrak{p}} = 1$ at every inert prime. Consider the decomposition of $yZ(\Lambda)$ into a product of powers of prime ideals in $Z(\Lambda)$. If $\mathcal{P}$ is over an inert prime then $\mathrm{ord}_{\mathcal{P}} y$ is even. At a ramified $\mathcal{P}_0$, $\mathrm{ord}_{\mathcal{P}_0} y$ is even because $y = \bar{y}$. Finally if $\mathcal{P}$ is over split then $\mathrm{ord}_{\mathcal{P}} y = \mathrm{ord}_{\bar{\mathcal{P}}} y$. Clearly we can choose a decomposition of $yZ(\Lambda)$ into the form $yZ(\Lambda) = A\bar{A}$.

Next suppose $yZ(\Lambda) = A\bar{A}$. Let $\mathcal{P}$ be over inert and $\pi = \bar{\pi}$ a local uniformizer for $\mathcal{O}(\mathcal{P})$. Then $A\cdot\mathcal{O}(\mathcal{P}) = (\pi^{\nu})$ and $yu = \pi^{2\nu}$, $u = \bar{u}$ for some $u \in \mathcal{O}(\mathfrak{p})^*$. This shows $\mathrm{ord}_{\mathfrak{p}}\, y \equiv 0 \pmod 2$ and hence $(y,\sigma)_{\mathfrak{p}} = 1$. Observe this is exactly what we should expect in light of (III.1.3).

Obviously the question is whether or not we could throw out fractional ideals and just use $Z(\Lambda)$ together with a unit in $Z(\Lambda + \lambda^{-1})^*$. In general the answer is no. To approach this question we shall have to introduce two short exact sequences of C_2-modules

$$1 \longrightarrow Z(\Lambda)^* \longrightarrow Q(\Lambda)^* \longrightarrow Q(\Lambda)^*/Z(\Lambda)^* \longrightarrow 1$$

and

$$1 \longrightarrow Q(\Lambda)^*/Z(\Lambda)^* \longrightarrow \mathcal{I} \longrightarrow \mathcal{C} \longrightarrow 1$$

where $\mathcal{I}$ is the group of fractional ideals and $\mathcal{C}$ is the ideal class group. The involution on $\mathcal{I}$ is $A \longrightarrow \hat{A}$ while on $Q(\Lambda)^*$ it is $x \longrightarrow \bar{x}^{-1}$. We should best do a bit of computation before running exact sequences together.

(2.4) Lemma: The cohomology group $H^2(C_2;Q(\Lambda)^*) = \{1\}$ by Hilbert Theorem 90. If n is composite, $H^2(C_2;Z(\Lambda)^*) = \{1\}$ while if $n = p^k$ then $H^2(C_2;Z(\Lambda)^*) \simeq C_2$.

Proof. A cohomology class in $H^2(C_2;Z(\Lambda)^*)$ arises from a unit for which $u = \hat{u} = \bar{u}^{-1}$. That is, $u\bar{u} = 1$. Thus u is a root of unity so $u = \pm\lambda^j$. If $\lambda^{2i} = \lambda^j$ then $\lambda^i\hat{\lambda}^i = \lambda^j$. If n is composite $\mu\bar{\mu} = \mu\bar{\mu}^{-1} = -1$. If $n = p^k$ there is no skew unit, but $-1 = \mu\hat{\mu}$ in $Q(\Lambda)^*$. □

(2.5) Lemma: The group $H^2(C_2;\mathcal{I}) = \{1\}$.

Proof. Suppose $A = \hat{A} = \bar{A}^{-1}$, then for any prime with $\bar{\mathcal{P}} = \mathcal{P}$ we must have $\mathrm{ord}_{\mathcal{P}}A = -\mathrm{ord}_{\mathcal{P}}A = 0$. If $\mathcal{P} \neq \bar{\mathcal{P}}$ then $\mathrm{ord}_{\mathcal{P}}A = -\mathrm{ord}_{\bar{\mathcal{P}}}A$. Clearly we have a B with $B\hat{B} = B\bar{B}^{-1} = A$. □

Now we get

$$\{1\} \longrightarrow H^2(C_2;\mathcal{C}) \longrightarrow H^3(C_2;Q(\Lambda)^*/Z(\Lambda)^*) \simeq H^1(C_2;Q(\Lambda)^*/Z(\Lambda)^*) .$$

This imbedding may be directly described as follows. If $A \sim \hat{A}$ then there is $y \in Q(\Lambda + \lambda^{-1})^*$ with $yZ(\Lambda) = A\bar{A}$. Now $c\ell(y) \in H^1(C_2;Q(\Lambda)^*/Z(\Lambda)^*)$ defines

$$1 \longrightarrow H^2(C_2;\mathcal{C}) \longrightarrow H^1(C_2;Q(\Lambda)^*/Z(\Lambda)^*) .$$

Note $c\ell(y) = 1$ if and only if there is $x \in Q(\Lambda)^*$ and $u \in Z(\Lambda)^*$ with $ux\hat{x}^{-1} = ux\bar{x} = y$. In other words if $H^2(C_2;\mathcal{C}) \neq \{1\}$ we cannot rid ourselves of these non-principal ideals even up to Witt equivalence.

(2.6) Lemma: The homomorphism $H^1(C_2;\mathcal{I}) \longrightarrow H^1(C_2;\mathcal{C})$ is trivial.

Proof. Let A be a fractional ideal for which $A\hat{A} = Z(\Lambda)$. Then $A = \bar{A}$. Consider $\mathfrak{p}$ inert only and $\mathcal{P}$ the overlying prime in $Z(\Lambda)$. By the realization of Hilbert symbols we can always find a $y \in Q(\Lambda + \lambda^{-1})^*$ such that at all inert $\mathfrak{p}$

$$(y,\sigma)_{\mathfrak{p}} = (-1)^{\mathrm{ord}_{\mathcal{P}}A} = (-1)^{\mathrm{ord}_{\mathfrak{p}}\, y} .$$

Then $\text{ord}_{\mathcal{P}} yA \equiv 0 \pmod 2$ if $\mathcal{P}$ is over an inert prime and if $\bar{\mathcal{P}} \neq \mathcal{P}$ then $\text{ord}_{\bar{\mathcal{P}}} yA = \text{ord}_{\mathcal{P}} yA$. If there is a ramified prime it is principal so we can assume $\text{ord}_{\mathcal{P}_0} yA = 0$. Then there is a fractional ideal B with $B\bar{B} = B\hat{B}^{-1} = yA$.

Thus we have a diagram

$$\begin{array}{ccccc} & & \{1\} & & \\ & & \downarrow & & \\ & & H^1(C_2;\mathscr{C}) & & \\ & & \downarrow & & \\ \{1\} = H^2(C_2;Q(\Lambda)^*) & \longrightarrow & H^2(C_2;Q(\Lambda)^*/Z(\Lambda)^*) & \longrightarrow & H^3(C_2;Z(\Lambda)^*) \\ & & & & \simeq H^1(C_2;Z(\Lambda)^*) \\ & & \downarrow & & \\ & & H^2(C_2;\mathscr{I}) = \{1\} \; . & & \end{array}$$

The resultant isomorphic embedding $\{1\} \to H^1(C_2;\mathscr{C}) \to H^1(C_2;Z(\Lambda)^*)$ has a direct description. Suppose $A\hat{A} = A\bar{A}^{-1} = xZ(\Lambda)$. Then $\bar{x}Z(\Lambda) = \bar{A}A^{-1} = x^{-1}Z(\Lambda)$. Hence $x\bar{x} = u \in Z(\Lambda)^*$ with $u\hat{u} = u\bar{u}^{-1} = 1$. Now $cl(A) \longrightarrow cl(u)$ is the embedding.

(2.7) Theorem: <u>There are two short exact sequences</u>

$$\{1\} \to H^2(C_2;\mathscr{C}) \to H^1(C_2;Q(\Lambda)^*/Z(\Lambda)^*) \to H^1(C_2;\mathscr{I}) \to \{1\}$$

and

$$\{1\} \to H^1(C_2;\mathscr{C}) \to H^1(Z(\Lambda)^*) \to \mathscr{N} = H^1(C_2;Q(\Lambda)^*) \; .$$

The second line accounts for units in $Z(\Lambda + \lambda^{-1})^*$ which are Hermitian squares, but not Hermitian squares of units in $Z(\Lambda)^*$. Abstractly $H^2(C_2;\mathscr{C}) \approx H^1(C_2;\mathscr{C})$ because $\mathscr{C}$ is finite. The smallest prime for which $H^2(C_2;\mathscr{C}) \neq \{1\}$ is 29. In that case $H^2(C_2;\mathscr{C}) \simeq C_2 \oplus C_2 \oplus C_2$.

There is an epimorphism

$$\mathscr{H}_0(Z(\Lambda)) \to H^2(C_2;\mathscr{C}) \longrightarrow \{1\} \; .$$

Briefly it is described as follows. Let (P,h) be a finitely generated projective

$Z(\Lambda)$ module with Hermitian innerproduct over $Z(\Lambda)$. Then $0 \quad P \quad P \otimes_{Z(\Lambda)} Q(\Lambda) = V$ and $\dim V = m$. Thus $0 \longrightarrow \Lambda^m P \longrightarrow \Lambda^m V = Q(\Lambda)$ and $\Lambda^n P$ inherits uniquely a Hermitian innerproduct ([M-H]). Thus $\Lambda^m P = A \subset Q(\Lambda)$ is a fractional ideal for which $A \sim \hat{A}$. Take $c\ell(A) \in H^2(C_2;\mathscr{C})$. If $N \subset P$ is a metabolizer then $0 \longrightarrow N \longrightarrow P \longrightarrow N^* \longrightarrow 0$ with $N^* = \mathrm{Hom}_{Z(\Lambda)}(N, Z(\Lambda))$ splits. So $\dim V = 2m$ and $\Lambda^{2m} P = \Lambda^m N \otimes_{Z(\Lambda)} \Lambda^m N^*$. If $B = \Lambda^m N$ then $\hat{B} = \Lambda^m N^*$ and $\Lambda^{2m} P = B\hat{B}$.

Another description is to use the projective class group $\tilde{K}_0(Z(\Lambda)) \simeq \mathscr{C}$. The involution on $\tilde{K}_0(Z(\Lambda))$ is $P \longrightarrow \mathrm{Hom}_{Z(\Lambda)}(P, Z(\Lambda))$. Then $\mathscr{H}_0(Z(\Lambda)) \longrightarrow H^2(C_2;\tilde{K}_0(Z(\Lambda))) \simeq H^2(C_2;\mathscr{C})$ can be defined by analogy with $W_0(D,S) \longrightarrow H^2(C_2; GR(D,S))$ introduced in section 1 of chapter I.

(2.8) <u>Exercise</u>: A Witt class in $\mathscr{H}_0(Z(\Lambda))$ contains a representative which is free as a $Z(\Lambda)$-module if and only if this class is in the kernel of $\mathscr{H}_0(Z(\Lambda)) \longrightarrow H^2(C_2;\mathscr{C}) \longrightarrow \{1\}$.

There is no problem in extending this analysis to $\mathscr{H}_0(D(N))$. For n composite or $n = p^k$ and $p|N$, $\mathscr{H}_2(D(N)) \simeq \mathscr{H}(D(N))$. For $n = p^k$ there are some interesting points about $\mathscr{H}_2(Z(\Lambda))$. We begin with a lemma that will also be used in another context.

(2.9) <u>Lemma</u>: <u>If</u> $C \subset Q(\Lambda)$ <u>is any fractional ideal then</u> $C/\mu C \simeq Z(\Lambda)/(\mu)$.

Proof. Let us assume $C \supset Z(\Lambda)$. For some $m \leq 0$, $C = \mu^m B$ with

$$(\mu) + B^{-1} = Z(\Lambda) .$$

Then $\mu B + Z(\Lambda) = B$ and

$$\mu^{m+1} B + (\mu^m) = \mu^m B = C .$$

Now $\mu^{m+1} B = \mu\mu^m B = \mu C$ so

$$\mu C + (\mu^m) = C .$$

The lemma follows. $\square$

We are specifically interested in the following situation. Let $A \subset Z(\Lambda)$ be an integral ideal and suppose there is a $y \in Q(\Lambda + \lambda^{-1})$ with $A\bar{A} = yZ(\Lambda)$. Then $A^{-1} \supset Z(\Lambda)$ and a Hermitian innerproduct on A^{-1} is given by

$$[\beta, \beta_1] = y\beta\bar{\beta}_1 \in Z(\Lambda) .$$

Now on $A^{-1}/\mu A^{-1}$ this induces

$$(\mathrm{cl}(\beta), \mathrm{cl}(\beta_1)) = \mathrm{cl}(y\beta\bar{\beta}_1) \in Z(\Lambda)/(\mu) .$$

This is a rank 1 symmetric form over the residue field $Z(\Lambda)/(\mu)$. In fact we can, using the formula

$$\mu A^{-1} + (\mu^m) = A^{-1} ,$$

give the form explicitly. Write $y = v(-\sigma)^{-m}$ with $v \in \mathcal{O}(\mathscr{P}_0)^*$. Then on $\mathcal{O}(\mathfrak{p}_0)/(-\sigma) = \mathcal{O}(\mathscr{P}_0)/(\mu) = Z(\Lambda)/(\mu)$ our form is $(x,y) \longrightarrow vxy$. Clearly $vx^2 = u$ has a solution if and only if

$$(y,\sigma)_{\mathfrak{p}_0} = (u,\sigma)_{\mathfrak{p}_0}$$

where $u \in \mathcal{O}(\mathfrak{p}_0)^*$.

(2.9) Lemma: *If $u \in Z(\Lambda + \lambda^{-1})$ then there is a $\beta \in A^{-1}$ with*

$$y\beta\bar{\beta} \equiv u \pmod{\sigma}$$

if and only if $(y,\sigma)_{\mathfrak{p}_0} = (u,\sigma)_{\mathfrak{p}_0}$.

Since $(y,\sigma)_{\mathfrak{p}} = (u,\sigma)_{\mathfrak{p}}$ at all other finite primes we can also express the condition by

$$\prod_{j=1}^{\phi(n)/2} (y^{-1}u,\sigma)_{\mathfrak{p}\,\infty_j} = 1 .$$

Now we are ready to construct some rank 2 forms in $\mathcal{H}_2(Z(\Lambda))$. Let $A \subset Z(\Lambda)$ be an integral ideal with $A\bar{A} = yZ(\Lambda)$, $y \in Q(\Lambda+\Lambda^{-1})^*$. Think of $Z(\Lambda) \oplus A \subset Q(\Lambda) \oplus Q(\Lambda)$. If

$$\begin{bmatrix} \mu\alpha & \beta \\ -\bar{\beta} & \mu\gamma \end{bmatrix}$$

with $\alpha, \gamma \in Q(\Lambda+\Lambda^{-1})$ is a skew-Hermitian matrix over $Q(\Lambda)$ with non-zero determinant then there is on $Q(\Lambda) \oplus Q(\Lambda)$ a skew-Hermitian innerproduct given by

$$H((v,w),(v_1,w_1)) = -\mu\alpha\, v\bar{v}_1 - \mu\gamma\, w\bar{w}_1 + \bar{\beta}\, v\bar{w}_1 - \beta\, w\bar{v}_1 .$$

We state the following without the proof, which is a routine verification

(2.10) __Lemma__: __The restriction of__ H __to__ $Z(\Lambda) \oplus A$ __is__ $Z(\Lambda)$__-valued if and only if__ $\alpha \in Z(\Lambda)$, $\beta \in A^{-1}$ and $y\gamma \in Z(\Lambda)$. __Furthermore this restriction is a skew-Hermitian innerproduct over__ $Z(\Lambda)$ __if and only if__

$$\sigma\alpha\gamma + \beta\bar{\beta} = y^{-1}u$$

__for some__ $u \in Z(\Lambda+\Lambda^{-1})^*$.

We could replace γ by $y^{-1}\gamma$, $\gamma \in Z(\Lambda)$. Then the determinent condition becomes

$$\sigma y^{-1}\alpha\gamma + \beta\bar{\beta} = y^{-1}u ,$$

and multiplying by y

$$\sigma\alpha\gamma + y\beta\bar{\beta} = u .$$

In III.2.9 we saw this can hold for a given $u \in Z(\lambda+\lambda^{-1})$ if and only if

$$(y,\sigma)_{\mathfrak{p}_0} = (u,\sigma)_{\mathfrak{p}_0} .$$

(2.11) Lemma: The homomorphism $Z(\lambda+\lambda^{-1})^* \longrightarrow (Z(\lambda)/(\mu))^*$ is an epimorphism.

This quotient identifies 1 with λ so $1+\lambda+\ldots+\lambda^j$, $1 \le j \le p-1$, which is in $Z(\lambda)^*$, is identified with $j\cdot 1$. Since every unit in $Z(\lambda)^*$ can be expressed as the product of a unit in $Z(\lambda+\lambda^{-1})^*$ with a root of unity the lemma follows. □

Recall we wrote $y = v(-\sigma)^{-m}$ where $v \in \mathcal{O}(\mathfrak{p}_0)^*$. We have just shown there is a unit $u \in Z(\lambda+\lambda^{-1})^*$ for which $vu = 1 \in \mathcal{O}(\mathcal{P}_0)/(\mu) = Z(\lambda)/(\mu)$. Now we conclude

(2.12) Theorem: If A is any fractional ideal with $A \sim \hat{A}$ then $Z(\lambda) \oplus A$ admits a skew-Hermitian $Z(\lambda)$-valued innerproduct.

3. The different over Q

If $Z(\lambda)$ is the ring of integers in the nth cyclotomic number field we want to introduce

$$0 \longrightarrow \mathcal{H}_2(Z(\lambda)) \longrightarrow W_0(Z,C_n)$$

$$0 \longrightarrow \mathcal{H}_0(Z(\lambda)) \longrightarrow W_2(Z,C_n) .$$

To do this we must introduce the different of $Q(\lambda)$ over Q (not over $Q(\lambda+\lambda^{-1})$). If we write $2m = \phi(n)$ and denote by $\phi(t)$ the nth cyclotomic polynomial then

$$t^{2m}\phi(t^{-1}) = \phi(t) .$$

Differentiating both sides with respect to t

$$\phi'(t) = 2mt^{2m-1}\phi(t^{-1}) - t^{2(m-1)}\phi'(t^{-1})$$

and, since $\phi(\lambda) = 0$, it follows

$$\phi'(\lambda) = -\lambda^{2(m-1)}\phi'(\lambda^{-1}) .$$

Hence if $\Delta = \lambda^{1-m}\phi'(\lambda) \in Z(\lambda)$ we find $\bar{\Delta} = -\Delta$ and (Δ) is the different ideal of the cyclotomic extension.

If V is a finitely generated projective $Z(\lambda)$-module then $\mathrm{Hom}_{Z(\lambda)}(V, Z(\lambda))$ is given the $Z(\lambda)$-module structure $(t\psi)(v) = \psi(\bar{t}v) = \bar{t}\psi(v)$ and $\mathrm{Hom}_Z(V,Z)$ the $Z(\lambda)$-module structure $(t\varphi)(v) = \varphi(\bar{t}v)$. We define a $Z(\lambda)$-module homomorphism

$$\mathrm{Tr}\colon \mathrm{Hom}_{Z(\lambda)}(V, Z(\lambda)) \longrightarrow \mathrm{Hom}_Z(V,Z)$$

by

$$(\mathrm{Tr}\,\psi)(v) = \mathrm{tr}(\Delta^{-1}\psi(v)) .$$

If $t \in Z(\lambda)$

$$(t(\mathrm{Tr}\,\psi))(v) = \mathrm{tr}(\Delta^{-1}\psi(\bar{t}v)) = \mathrm{tr}(\Delta^{-1}\bar{t}\psi(v)) = \mathrm{tr}(\Delta^{-1}(t\psi)(v)) .$$

(3.1) Lemma: For any finitely generated projective $Z(\lambda)$-module

$$\mathrm{Tr}\colon \mathrm{Hom}_{Z(\lambda)}(V, Z(\lambda)) \simeq \mathrm{Hom}_Z(V,Z) .$$

Proof. By definition of the different this is true if $V = Z(\lambda)$. Thus it is true for V free and hence for V projective. □

Now by taking $Z = D$ and $Z(\lambda) = S$ we have from Chapter I the Witt groups $W_0(Z, Z(\lambda))$ and $W_2(Z, Z(\lambda))$. By analogy with I.2.7 we find

(3.2) Theorem: There are isomorphisms

$$\mathcal{H}_2(Z(\lambda)) \simeq W_0(Z, Z(\lambda))$$
$$\mathcal{H}_0(Z(\lambda)) \simeq W_2(Z, Z(\lambda)) .$$

The fact that $\bar{\Delta} = -\Delta$ produces the parity interchange.

We can even do the following. Consider only divisors $m|n$ with $m > 1$. Let $Z(\lambda_m)$, $\lambda_m = \exp 2\pi i/m$, be the integers in the mth cyclotomic extension. Then there is

$$0 \longrightarrow W(Z) \oplus \sum_{m|n} \mathcal{H}_2(Z(\lambda_m)) \longrightarrow W_0(Z, C_n)$$

$$0 \longrightarrow \sum_{m|n} \mathcal{H}_0(Z(\lambda_m)) \longrightarrow W_2(Z, C_n) .$$

Let us talk about this a bit. For each divisor, including $m = 1$, this time, let $\Phi_m(t)$ be the mth cyclotomic polynomial. If $a_0 = \pm 1$ is the constant term then

$$\Phi_m(t^{-1}) = a_0 t^{-\varphi(m)} \Phi_m(t) .$$

Let $C_m(t)$ be the polynomial for which $\Phi_m(t) C_m(t) = x^n - 1$. If $d(m)$ is the degree of $C_m(t)$ then

$$C_m(t^{-1}) = -a_0 \, t^{-d(m)} C_m(t) .$$

Now consider (T, V, b) where (V, b) is an innerproduct space over Z and (T, V) is an isometry for which $T^n = Id$. For each $m|n$ let $V_m \subset V$ be the T-invariant subspace

$$V_m = \{v | v \in V, \Phi_m(T)v = 0\} .$$

The minimal polynomial of $T|V_m$, if $V_m \neq \{0\}$, is $\Phi_m(t)$ and the period of this restriction is exactly m. Note $C_m(T)V \subset V_m$ with finite index.

(3.3) <u>Lemma</u>: <u>If</u> $m \neq m_1$ <u>then</u> $V_m \subset (V_{m_1})^{\perp}$.

Proof. If $v \in V_m$ then there is $N > 0$ for which $Nv \in C_m(T)V$. Since

$\Phi_{m_1}(t)$ is a factor for $C_m(T)$ we have $Nv = \Phi_{m_1}(T)w$. Thus if $v_1 \in V_{m_1}$, $N\beta(v,v_1) = \beta(Nv,v_1) = \beta(\Phi_{m_1}(T)w,v) = \beta(w,\Phi_{m_1}(T^{-1})v_1) = \beta(w,a_0T^{-\varphi(m_1)}\Phi_{m_1}(T)v_1) = 0$. □

Obviously $V_m \cap V_{m_1} = \{0\}$ if $m \neq m_1$. Look at $V_1 \oplus \ldots \oplus V_n$. This is T-invariant and has finite index in V. Each V_m is a module over $Z[C_m]/(\Phi_m(T)) = Z(\Lambda_m)$. The restriction of b to V_m has a non-zero determinant (which is a unit in $Z(1/n)$).

(3.4) Lemma: Let $N \subset V$ be a T-invariant metabolizer. If $N_m = N \cap V_m$ then $N_m^{\perp} = N_m$ where orthogonal complement is taken in V_m with respect to $b|V_m$.

Proof. Suppose $v \in V_m$ is in the complement of N_m. Now $C_m(T)N \subset N_m$, hence $0 = b(v, C_m(T)N) = b(C_m(T^{-1})v, N) = b(-a_0T^{-d(m)}C_m(T)v, N) = a_0b(C_m(T)v, T^{d(m)}N = N)$. Thus $C_m(T)v \in N^{\perp} = N$. If $0 \neq k \in Z$ is the resultant of $\Phi_m(t)$, $C_m(t)$ then there are integral polynomials $r(t)$, $s(t)$ with $r(t)\Phi_m(t) + s(t)C_m(t) = k$.

Since $\Phi_m(T)v = 0$ we find

$$S(T)C_m(t)v = kv\ .$$

As N is T-invariant, $kv \in N$ also. Of course N is a summand and hence $v \in N$. That is, $v \in V_m \cap N = N_m$.

It now follows that

$$0 \longrightarrow W(Z) \oplus \sum \mathcal{H}_s(Z(\Lambda_m)) \longrightarrow W_0(Z, C_n)$$
$$0 \longrightarrow \sum \mathcal{H}(Z(\Lambda_m)) \longrightarrow W_2(Z, C_n)$$

are monomorphisms as indicated.

Surely this looks like the start of some sort of exact sequence. This will be the subject of our next section.

4. Coupling Invariants

The subject in this section is the discussion of an elegant idea due to

Stoltzfus ([Sf]).

First let us define $W_0(Q/Z,\Phi_m)$. Consider (T,A,β) where (A,β) is a Q/Z-valued symmetric innerproduct on a finite abelian group and $T\colon A \longrightarrow A$ is an isometry for which $\Phi_m(T) = 0$. We make up the usual Witt group and call it $W_0(Q/Z,\Phi_m)$ $= W_0(Q/Z,Z(\lambda_m))$. For a prime p we make the analogous definition for $W_0(F_p,\Phi_m)$. By appealing to anisotropics

$$\sum_p W_0(F_p,\Phi_m) \simeq W_0(Q/Z,\Phi_m) \ .$$

To analyze $W_0(F_p,\Phi_m)$ further it is necessary to consider the factorization of $\Phi_m(t)$ in $F_p[t]$. This is in line with $W_0(Q/Z,S)$ in Chapter I. This is postponed to later.

The idea is to define a boundary homomorphism

$$\text{cup}\colon W_0(Z,C_n) \longrightarrow \sum_{m|n} W_0(Q/Z,\Phi_m)$$

which is called the coupling invariant. Given an integral (T,V,b) we found the subspaces V_m for all $m|d$. Take $W = V \otimes_Z Q$ and consider (T,W,b) a rational innerproduce space with isometry. This time, because $Q[t]$, unlike $Z[t]$, is a principal ideal domain we do get an orthogonal decomposition $W = W_1 \oplus \ldots \oplus W_n$.

Note that $V_m \subset W_m$ is a T-invariant integral lattice for (W_m,b_m). Then $V_m^{\#}/V_m$ has a Q/Z form and an induced isometry, still called T, for which $\Phi_m(T) = 0$. In this manner we receive $\text{cup}\langle T,V,b\rangle$ in $\sum_{m|n} W_0(Q/Z,\Phi_m)$. From III.3.4 it follows that cup is well defined.

What does $\text{cup}\langle T,V,b\rangle = 0$ mean? For a kernel element we have T-invariant self-dual lattices

$$V_m \subset L_m \subset W_m \ .$$

Putting $(T,L,b) = \sum_{m|n} (T,L_m,b_m)$ we get an orthogonally decomposed isometry over Z. Thus

$$\langle T,V,b\rangle = \sum \langle T,L_m,b_m\rangle$$

and $\langle T,L_m,b_m\rangle$ is in the image of $W_0(Z,Z(\Lambda_m)) \longrightarrow W_0(Z,C_n)$. So far we have shown

$$0 \to W(Z) \oplus \sum \mathcal{H}(Z(\Lambda_m)) \to W_0(Z,C_n) \xrightarrow{\text{cup}} \sum W_0(Q/Z,\Phi_m)$$

We can go on from here to $W_0(Q/Z,\Phi_m) \longrightarrow W_0(Q/Z,C_n)$. Thus we get

$$\sum_{m|n} W_0(Q/Z,\Phi_m) \xrightarrow{\Gamma} W_0(Q/Z,C_n)$$

simply by adding up in $W_0(Q/Z,C_n)$. The composition $\Gamma \circ \text{cup}\colon W_0(Z,C_n) \longrightarrow W_0(Q/Z,C_n)$ is trivial because this composition would be nothing but the restriction of $\partial\colon W_0(Q,C_n) \longrightarrow W_0(Q/Z,C_n)$ to the image of

$$0 \longrightarrow W_0(Z,C_n) \longrightarrow W_0(Q,C_n) \ .$$

We would like to show $\text{im}(\text{cup}) = \ker(\Gamma)$. Actually we can extend cup to $W_0(Q,C_n)$. Given a rational (T,W,b) we split up orthogonally into $\sum_{m|n} (T,W_m,b_m)$. In W_m choose a T-invariant integral lattice $L_m \subset W_m$ with respect to b_m. Take $L_m^{\#} \subset W_m$ so that we receive $\langle T,L_m^{\#}/L,\beta_m\rangle \in W_0(Q/Z,\Phi_m)$.

Surely $L = \sum L_m$ is an integral lattice for (T,W,b). It must be observed that $L^{\#} = \sum_{m|n} L_n^{\#}$. If $v \in L^{\#}$ then we can find $v_m \in L_m^{\#} \subset W_m$ with $v - v_m \in W_m^{\perp}$ and $v_m \in \cap W_{m'}^{\perp}$, $m' \neq m$. Then $v - \sum_{n|m} v_m$ lies in $W^{\perp} = \{0\}$.

Hence $\Gamma \circ \text{cup} = \partial$, where cup is extended to $W_0(Q,C_n)$. If $\partial\langle T,W,b\rangle = 0$ then there is a self dual integral lattice $\sum L_m \subset \mathcal{L} \subset W$. Then $\text{cup}\langle T,W,b\rangle = \text{cup}\langle T,\mathcal{L},b\rangle$.

(4.1) Theorem: There is an exact sequence

$$0 \to W(Z) \oplus \sum \mathcal{H}_s(Z(\lambda_m)) \to W_0(Z,C_n) \xrightarrow{\text{cup}} \sum_{m|n} W_0(Q/Z,\Phi_m) \xrightarrow{\Gamma} W_0(Q/Z,C_n) .$$

For the skew-symmetric case we would make similar definitions and arrive at

$$0 \to \sum \mathcal{H}(Z(\lambda)) \to W_2(Z,C_n) \xrightarrow{\text{cup}} \sum W_2(Q/Z,\Phi_m) \xrightarrow{\Gamma} W_2(Q/Z,C_n) .$$

5. A little more algebra

This section is used as background material for the computation of $W_*(Z,C_n)$. Let $m \geq 1$, p be odd integers, p a prime and $(m,p) = 1$. There is the cyclotomic number field $Q(\lambda_m)$ with $\lambda_m = \exp(2\pi i/m)$. We allow $m = 1$. There is the ring of integers $Z(\lambda_m) \subset Q(\lambda_m)$, and we adjoin $1/mp$. So denote the resulting ring by $Z(\lambda_m/mp)$. This is the intersection of those local rings of integers corresponding to primes in $Z(\lambda_m)$ which do not divide mp. Then $Z(\lambda_m/mp)$ is still a Dedekind domain and in fact is the integral closure in $Q(\lambda_m)$ of $Z(1/mp)$.

Now for $r \geq 1$ there is $Q(\lambda_{mp^r})$ which we think of as a relative abelian extension of $Q(\lambda_m)$ with degree $\varphi(p^r)$. In fact the minimal polynomial of λ_{p^r} over $Q(\lambda_m)$ is still just the cyclotomic polynomial $\Phi_{p^r}(t)$. As a consequence the relative different of this extension is a unit in $Z(\lambda_{mp^r}/mp)$.

We put $S = Z(\lambda_{mp^r}/mp)$ and let complex conjugation act as the involution. Of course $Z(\lambda_m/mp) \subset S$ and is stable with respect to conjugation. We would like to compute $\mathcal{H}_*(Z(\lambda_m/mp),S)$. This is the Hermitian analogue of $W_*(D,S)$. We consider triples (S,V,h) wherein

a) (S,V) is an S-module which is finitely generated and projective as a $Z(\lambda_m/mp)$-module

b) (V,h) is a (skew)-Hermitian innerproduct space structure over $Z(\lambda_m/mp)$

c) $h(sv,w) \equiv h(v,\bar{s}w)$, all $s \in S$.

The reader can make up the canonical Witt class definition. We state by analogy

(5.1) Lemma: The relative trace homomorphism induces an isomorphism

$$\mathcal{H}_*(Z(\Lambda_{mp^r}/mp)) \simeq \mathcal{H}_*(Z(\Lambda_m/mp),S) .$$

If $(W,\langle\, ,\,\rangle)$ is a (skew)-Hermitian innerproduct space over $Z(\Lambda_{mp^r}/mp)$ we put, using relative trace,

$$h(v,w) = tr\langle v,w\rangle$$

to obtain (S,W,h). We again remark that the relative different is a unit in $Z(\Lambda_{mp^r}/mp)$.

There is one other issue to be discussed, and that is the relation of $\mathcal{H}_*(Z(\Lambda_m/m))$ to $\mathcal{H}_*(Z(\Lambda_m/mp))$. If $D = Z(\Lambda_m/m)$ then we could have set up

$$0 \longrightarrow \mathcal{H}_*(D) \longrightarrow \mathcal{H}_*(Q(\Lambda_m)) \xrightarrow{\partial} \mathcal{H}_*(E/D) \ , \ E = Q(\Lambda_m) .$$

Now $\mathcal{H}_*(E/D)$ splits up into a sum $\sum \mathcal{H}_*(E/\mathcal{O}(\mathcal{P}))$ which runs over the local rings corresponding to primes $\mathcal{P}$ in $Z(\Lambda_m/m)$ for which $\bar{\mathcal{P}} = \mathcal{P}$. Now ∂ should be thought of as very local. Indeed it splits up into $\partial_{\mathcal{P}}: \mathcal{H}_*(E) \longrightarrow \mathcal{H}_*(E/\mathcal{O}(\mathcal{P}))$.

Thus to relate $\mathcal{H}(Z(\Lambda_m/m))$ to $\mathcal{H}(Z(\Lambda_m/mp)$ we would pick the primes $\mathcal{P}_1,\ldots,\mathcal{P}_r$ in $Z(\Lambda_m/m)$ for which $p = \mathcal{P}_1 \ldots \mathcal{P}_r$. The ramification indices are all 1 since $(p,m) = 1$. Either all $\mathcal{P}_j$ are conjugation invariant or none are. If none are then

$$\mathcal{H}_*(Z(\Lambda_m/m)) \simeq \mathcal{H}(Z(\Lambda_m/mp))$$

while if they all are then

$$0 \to \mathcal{H}_*(Z(\Lambda_m/m)) \to \mathcal{H}_*(Z(\Lambda_m/mp)) \to \sum_{j=1}^{r} \mathcal{H}_*(\mathcal{O}(\mathcal{P}_j)/m(\mathcal{P}_j)) .$$

In this last the induced involution on the residue field $\mathcal{O}(\mathcal{P}_j)/m(\mathcal{P}_j)$ is non-trivial.

6. Computing $W_*(Z, C_n)$

We need an inductive step first. Suppose $\ell = p^r k$ with $(k,p) = 1$. We want to compute $\mathcal{H}_*(Z\Lambda_m/mp), C_\ell)$ and compare with $\mathcal{H}_*(Z\Lambda_m/m), C_\ell)$. These are Witt groups of (skew)-Hermitian innerproducts carrying C_ℓ as a group of isometries. We also assume $(m,\ell) = 1$.

Look at a (C_ℓ, V, k). We may think of $C_\ell = C_{p^r} \times C_k$, and take T to be the preferred generator of C_{p^r}. Then we have $\Delta = T^{\frac{p^r-1}{2}} - T^{\frac{p^r+1}{2}}$ and for $1 \leq s \leq r$ we introduce operators

$$\Psi_s(T) = T^{\frac{-\varphi(p^s)}{2}} \Phi_{p^s}(T) .$$

The rational behind the latter is the identity $t^{-\varphi(p^s)}\Phi_{p^s}(t^{-1}) = \Phi_{p^s}(t)$ for the cyclotomic polynomials. Thus $\Psi_s(T) = \Psi_s(T^{-1})$.

Now V is a $Z\Lambda_m/mp)$-module so if $V_0 = \{v | \Delta v = 0\}$ and for $1 \leq s \leq r$ $V_s = \{v | \Psi_s v = 0\}$ then we receive an orthogonal direct sum splitting

$$V = V_0 \oplus V_1 \oplus \dots \oplus V_r .$$

Each V_j is still an innerproduct space over $Z\Lambda_m/mp)$ on which C_ℓ acts. All we have done is to arrange Δ to be skew-adjoint while each Ψ_j is self-adjoint. Now

$$t^{p^r} - 1 = (t-1)\Phi_p(t) \dots \Phi_{p^r}(t)$$

and in $Z(1/p)[t]$ these polynomials are pairwise relatively prime.

Obviously C_{p^r} acts trivially on V_0 so we obtain $\langle C_k, V_0, k\rangle \in \mathcal{H}_*(Z\Lambda_m/mp), C_k)$. In fact using $C_\ell = C_{p^r} \times C_k$ we find $\mathcal{H}_*(Z\Lambda_m/mp), C_k)$ now splits off from

$\mathcal{H}_*(Z(\Lambda_m/mp), C_\ell)$ as a direct summand.

Next on each V_s, $1 \leq s \leq r$, the T acts as multiplication by λ_{p^s}; in other words V_s is a $Z(\Lambda_{mp^s}/mp)$ module. To be absolutely precise it is a $Z(\Lambda_{mp^s}/mp)[C_k]$-module. But if the reader wishes to pursue this viewpoint he must think of the involution on this group ring as complex conjugation on the coefficient ring together with inversion of the group elements.

We apply therefore

$$\mathcal{H}_*(Z(\Lambda_{mp^s}/mp), C_k) \simeq \mathcal{H}_*(Z(\Lambda_m/mp), Z(\Lambda_{mp^s}/mp)[C_k])$$

by the relative trace as in Lemma III.3.2. Thus we arrive at a direct sum splitting

$$0 \to \mathcal{H}_*(Z(\Lambda_m/mp), C_k) \to \mathcal{H}_*(Z(\Lambda_m/mp), C_\ell) \to \bigoplus_1^r \mathcal{H}_*(Z(\Lambda_{mp^s}/mp), C_k) \to 0 .$$

Next return to the prime ideals $\mathcal{P}_1, \ldots, \mathcal{P}_r$ in $Z(\Lambda_m/m)$ over the rational prime p. Suppose $\bar{\mathcal{P}}_j = \mathcal{P}_j$ for all j. Then we arrive at the induction diagram for computing

$$\begin{array}{ccccccc}
0 \to \mathcal{H}_*(Z(\Lambda_m/m), C_k) & \to & \mathcal{H}_*(Z(\Lambda_m/mp), C_k) & \to & \sum \mathcal{H}_*(\mathcal{O}(\mathcal{P}_j)/m(\mathcal{P}_j), C_k) \\
\downarrow & & \downarrow & & \downarrow \simeq \\
0 \to \mathcal{H}_*(Z(\Lambda_m/m), C_\ell) & \to & \mathcal{H}_*(Z(\Lambda_m/mp), C_\ell) & \to & \sum \mathcal{H}_*(\mathcal{O}(\mathcal{P}_j)/m(\mathcal{P}_j), C_\ell) \\
 & & \downarrow & & \\
 & & \bigoplus_1^r \mathcal{H}_*(Z(\Lambda_{mp^s}/mp), C_k) & & \\
 & & \downarrow & & \\
 & & 0 & &
\end{array}$$

The middle column is a direct sum splitting. For the assertion

$$\mathcal{H}_*(\mathcal{O}(\mathcal{P}_j)/m(\mathcal{P}_j), C_k) \simeq \mathcal{H}_*(\mathcal{O}(\mathcal{P}_j)/m(\mathcal{P}_j), C_\ell)$$

we observe only that in an anisotropic representative over a field of characteristic p the subgroup C_{p^r} acts trivially.

Incidentally, if $\mathcal{P}_j \neq \bar{\mathcal{P}}_j$ we find

$$\mathcal{H}_*(Z(\Lambda_m/m), C_k) \simeq \mathcal{H}_*(Z(\Lambda_m/mp), C_k)$$
$$\mathcal{H}_*(Z(\Lambda_m/m), C_\ell) \simeq \mathcal{H}_*(Z(\Lambda_m/mp), C_\ell) .$$

In either case we arrive at the conclusion

$$\mathcal{H}_*(Z(\Lambda_m/m), C_\ell) \simeq \mathcal{H}_*(Z(\Lambda_m/m; C_k) \oplus \bigoplus_1^r \mathcal{H}_*(\Lambda_{mp^s}/mp), C_k) .$$

Applied iteratively this shows finally $\mathcal{H}_*(Z(\Lambda_m/m), C_\ell) = \sum_{k|\ell} \mathcal{H}_*(\Lambda_{mk}/mk)$ where the sum is taken over all divisors of ℓ.

(6.1) <u>Theorem</u>: <u>For C_n a cyclic group of odd order there is an (unnatural) additive isomorphism</u>

$$W_*(Z, C_n) \simeq \bigoplus_{k|n} \mathcal{H}_*(Z(\Lambda_k/k))$$

<u>with</u> $\mathcal{H}_*(Z(\Lambda_1/1)) = W_*(Z)$.

If $n = p^r$ we apply the foregoing with $m = 1$ and $k = 1$ to conclude $W_*(Z, C_{p^k})$ has no torsion.

We shall remind the reader how the torsion subgroup of $\mathcal{H}_*(\Lambda_k/k)$ is computed. Let $n(k)$ be the number of prime ideals in $Z(\Lambda_k + \lambda_k^{-1})$ which divide k and which are inert with respect to the relative quadratic extension $Q(\Lambda_k)/Q(\Lambda_k + \lambda_k^{-1})$, then the torsion subgroup of $\mathcal{H}_*(Z(\Lambda_k/k))$ is an elementary abelian 2-group of order $2^{n(k)-1}$. This is only of interest when k is composite and then $n(k)$ is determined as follows. For each prime p dividing k define an integer $r(p)$ as follows. Write $k = p^j m$ with $(p,m) = 1$, then

> $r(p) = 0$ if -1 is not contained in the subgroup of
> Z_m^* generated by p.
>
> otherwise $r(p)f = \varphi(m)$ where f is the smallest
> positive integer such that $p^f \equiv 1 \pmod{m}$.

Now $n(k)$ is the sum of the $r(p)$ taken over all primes dividing k.

Since this will enable the reader to look at a few elementary cases let us explain how it works. For the composite case, as mentioned earlier in this chapter, no finite prime of $Z(\lambda_k + \lambda_k^{-1})$ ramifies in $Q(\lambda_k)/Q(\lambda_k + \lambda_k^{-1})$. That is, every finite prime in $Z(\lambda_k + \lambda_k^{-1})$ is either split or inert. Thus $n(k)$ can be found by considering for each prime $p|k$ the number of prime ideals in $Z(\lambda_k)$ which divide p and for which $\bar{\mathcal{P}} = \mathcal{P}$. This will be the number $r(p)$. We write $k = p^j m$ with $(p,m) = 1$. Then first there will be $Q(\lambda_m)/Q$ of degree $\varphi(m)$. Here p does not ramify and so p factors into a product

$$\mathfrak{Q}_1 \cdots \mathfrak{Q}_r$$

of distinct prime ideals. The r is found from $rf = \varphi(m)$ with f the smallest positive number so that $p^f \equiv 1 \bmod (m)$. Now either all $\mathfrak{Q}_i$ are stable under conjugation or none are. As we shall later see $\mathfrak{Q}_i = \bar{\mathfrak{Q}}_i$ if and only if -1 is contained in the subgroup of Z_m^* generated by p. Now what will happen in the relative extension $Q(\lambda_{p^j m})/Q(\lambda_m)$? Each $\mathfrak{Q}_i$ will ramify with $e = \varphi(p^j)$. That is, $\mathcal{P}_i^{\varphi(p^j)} = \mathfrak{Q}_i$. So over each $\mathfrak{Q}_i$ there is just one $\mathcal{P}_i \subset Z(\lambda_k)$. Obviously $\bar{\mathcal{P}}_i = \mathcal{P}_i$ if and only if $\bar{\mathfrak{Q}}_i = \mathfrak{Q}_i$. Hence the counting process for $n(k)$ works.

Look at C_{15}. Here

$$W_0(Z, C_{15}) = W(Z) \oplus \mathcal{H}(Z(\lambda_5/5) \oplus \mathcal{H}(Z(\lambda_3/3)) \oplus \mathcal{H}(Z(\lambda_{15}/15)) .$$

Then $n(k)$ is computed as follows. First $5 \bmod 3$ is -1 so $r(5) = 1$, and $3^2 = -1 \bmod 5$ so $r(3) = 1$ also. Hence $n(15) = 2$ and $W(Z, C_{15})$ contains a non-trivial element of order 2. By contrast, $W_0(Z, C_{21})$ is without torsion. First $7 \equiv 1 \pmod 3$ so $r(7) = 0$ while $3^3 = -1 \pmod 7$, thus $r(3) = 1$ and $n(21) = 1$. It turns out that $n(55) = 0$.

As a challenge to the reader we propose an exercise. On $Z(\lambda_{15})$ define an innerproduct with values in Z by

$$(x,y) = \frac{1}{15} \operatorname{Tr}((\lambda_{15} + \lambda_{15}^{-1})(\lambda_{15}^{6} - \lambda_{15}^{-6})(\lambda_{15}^{5} - \lambda_{15}^{-5}) x\bar{y}) .$$

This is a unimodular C_{15} invariant innerproduct. Let $\mathcal{R}_k$ be the regular representation of C_k with its usual innerproduct. Show

$$[Z(\lambda_{15}), (,)] + \mathcal{R}_{15} - \mathcal{R}_5 - \mathcal{R}_3 + \mathcal{R}_1$$

represents the torsion class in $W_0(Z, C_{15})$.

Chapter IV

$W_*(Q, C_n)$ and Dress Induction

In this chapter we shall investigate the structure of $W_*(Q, C_n)$ for n an odd integer. In particular in section 3 we shall describe the invariants necessary to determine an element of $W_*(Q, C_n)$. Some of these invariants will have values in cohomology groups associated with the Grothendieck group of representations of C_n over a finite field. (See Note at end of Chapter VI.) The first two sections are devoted to a review of the basic properties of $GR(D, \pi)$, applications, Euler characteristics, and the relations between representations of cyclic groups and the characteristic polynomials. In section 4 we briefly indicate how these results extend to general odd order groups by appealing to Dress' induction theorems which we quote without proof.

1. The Grothendieck group of representations

We open this section with a brief reminder. If R is a commutative ring, π

a group, M, L both $R(\pi)$-modules then $\mathrm{Ext}_R^n(M,L)$ receives naturally the structure of an $R(\pi)$-module. Take a free $R(\pi)$-resolution of M,

$$\xrightarrow{\partial_{n+1}} F_n \to \dots \to F_1 \xrightarrow{\partial_1} F_0 \xrightarrow{\varepsilon} M \longrightarrow 0$$

and give to $C_R^n(M,L) = \mathrm{Hom}_R(F_n,L)$ the $R(\pi)$-module structure

$$(g\varphi)(x) = g\varphi(g^{-1}x))$$

where $g \in \pi$, $\varphi \in \mathrm{Hom}_R(F_n,L)$ and $x \in F_n$. This structure is passed on to $\mathrm{Ext}_R^n(M,L)$.

Now let D denote the integers or a field and π is a finite group. Let $\mathcal{F}$ be the free abelian group generated by finitely generated $D(\pi)$-modules (π,M). To each short exact sequence of finitely generated $D(\pi)$-modules

$$0 \longrightarrow (\pi,M_1) \xrightarrow{\alpha} (\pi,M) \xrightarrow{\beta} (\pi,M_2) \longrightarrow 0$$

there is associated a relation

$$(\pi,M) - (\pi,M_1) - (\pi,M_2)$$

in $\mathcal{F}$. If $\mathcal{R} \subset \mathcal{F}$ is the subgroup generated by all such relations, then the quotient $\mathcal{F}/\mathcal{R}$ is canonically isomorphic to the Grothendieck representation group $GR(D,\pi)$ introduced in section 1 of Chapter I (Swan [Sw 1,2]). Swan has studied this group extensively. We shall follow his viewpoint and quote his results.

We are particularly interested in the endomorphism of $\mathcal{F}$ given by

$$(\pi,M)^0 \equiv (\pi,\mathrm{Ext}_D^0(M,D)) - (\pi,\mathrm{Ext}_D^1(M,D)) \ .$$

(For a field of course the second term is absent.) This induces an endomorphism on

$GR(D,\pi)$. Simply observe that a short exact sequence

$$0 \longrightarrow (\pi,M_1) \xrightarrow{\alpha} (\pi,M) \xrightarrow{\beta} (\pi,M_2) \longrightarrow 0$$

induces a six term exact sequence of $D(\pi)$-modules:

$$\begin{aligned} 0 \longrightarrow \mathrm{Ext}_D^0(M_2,D) &\longrightarrow \mathrm{Ext}_D^0(M,D) \longrightarrow \mathrm{Ext}_D^0(M_1,D) \longrightarrow \\ \mathrm{Ext}_D^1(M_2,D) &\longrightarrow \mathrm{Ext}_D^1(M,D) \longrightarrow \mathrm{Ext}_D^1(M_1,D) \longrightarrow 0 \end{aligned}$$

so that in $GR(D,\pi)$:

$$\begin{aligned} \mathrm{Ext}_D^0(M,D) - \mathrm{Ext}_D^1(M,D) &= \mathrm{Ext}_D^0(M_1,D) - \mathrm{Ext}_0^1(M_1,D) \\ &+ \mathrm{Ext}_D^0(M_2,D) - \mathrm{Ext}_D^1(M_2,D) \ . \end{aligned}$$

For D a field it is obvious this endomorphism agrees with the involution on $GR(D,\pi)$ in Chapter I. In fact this is also the case for Z. If (π,M) is a finitely generated $Z(\pi)$-module let $T(m) \subset M$ be the Z-torsion submodule. The quotient $M/T(M)$ is also a $Z(\pi)$-module and a free abelian group. In $GR(Z,\pi)$ we find

$$(\pi,M) = (\pi,M/T(M)) + (\pi,T(M)) \ .$$

Also we check that

$$\mathrm{Ext}_Z^1(M,Z) \simeq \mathrm{Ext}_Z^1(T(M),Z) \simeq \mathrm{Hom}_Z(T(M),Q(Z)) \ .$$

Thus we can write in $GR(Z,\pi)$

$$(\pi,M)^0 = (\pi,\mathrm{Hom}_Z(M/T(M),Z)) - (\pi,\mathrm{Hom}_Z(T(M),Q/Z)) \ .$$

Since

$$(\pi, M/T(M)) \simeq (\pi, \mathrm{Hom}_Z(\mathrm{Hom}_Z(M,Z),Z))$$

$$(\pi, T(M)) \simeq (\pi, \mathrm{Hom}_Z(\mathrm{Hom}_Z(T(M),Q/Z),Q/Z))$$

it follows that we do have a "duality" involution on $GR(D,\pi)$. As in Chapter I, we think of this as an action of C_2 so that $H^*(C_2;GR(D,\pi))$ is defined.

For topological reasons it is the image of the Burnside ring of π in $GR(D,\pi)$ which is of interest. We recall $\mathrm{Burn}(\pi)$ is defined as follows. Consider all right actions (S,π) in which S is a finite, non-empty set. There is the free abelian group $\mathcal{B}$ generated by π-equivariant isomorphisms of such actions. If $(S,\pi) = (S_j \cup S_2, \pi)$, the disjoint union of two non-empty π-invariant subsets, then $\mathcal{B}_0$ is the subgroup generated by the relations $(S,\pi) - (S_1,\pi) - (S_2,\pi)$.

(1.1) <u>Definition</u>: <u>The Burnside ring of</u> π <u>is</u>

$$\mathrm{Burn}(\pi) = \mathcal{B}/\mathcal{B}_0 \ .$$

The multiplication is given by $(S_1,\pi) \times (S_2,\pi) = (S_1 \times S_2, \pi)$ where π acts diagonally on the product.

To define $\mathrm{Burn}(\pi) \longrightarrow GR(D,\pi)$ we associate to (S,π) the left $D(\pi)$-module $(\pi, D(S))$ where $D(S)$ is the free D-module of all functions on S into D and $(g\cdot f)(s) \equiv f(sg)$ for $g \in \pi$ defines the $D(\pi)$-module structure. In addition on $D(S)$ there is the innerproduct

$$b(f_1,f_2) = \sum_S f_1(s)f_2(s) \ .$$

In this π acts as a group of isometries and the adjoint

$$\mathrm{adj}_b : (\pi, D(S)) \longrightarrow (\pi, \mathrm{Hom}_D(D(S),D))$$

is an isomorphism. So we in fact have

$$\mathrm{Burn}(\pi)/2\ \mathrm{Burn}(\pi) \longrightarrow H^2(C_2;\mathrm{GR}(D,\pi)) .$$

Let us talk about a topological situation. Suppose (K,π) is a right action of π on a finite m-dimensional simplicial complex. Each group element is a simplical map and we may assume by barycentrically subdividing that if a simplex is mapped onto itself by some group element then it is pointwise fixed under that group element.

There is associated an Euler characteristic

$$\chi(K,\pi) = \sum_{n=0}^{m} (-1)^n (\pi, H^n(K;D)) \in \mathrm{GR}(D,\pi) .$$

We claim this element lies in the image of the Burnside ring. Clearly π acts on each cochain group $C^n(K;D)$ by a permutation representation. For each n there are two short exact sequences

$$0 \longrightarrow (\pi, Z^n(K;D)) \xrightarrow{i} (\pi, C^n(K;D)) \xrightarrow{\delta} (\pi, B^{n+1}(K;D)) \longrightarrow 0$$

and

$$0 \longrightarrow (\pi, B^n(K;D)) \xrightarrow{i} (\pi, Z^n(K;D)) \longrightarrow (\pi, H^n(K;D)) \longrightarrow 0 .$$

Hence in $\mathrm{GR}(D,\pi)$

$$(\pi, C^n(K;D)) = (\pi, H^n(K;D)) + (\pi, B^n(K;D)) + (\pi, B^{n+1}(K;D)) .$$

Since $B^0(K;D) = 0$ we find

$$\chi(K,\pi) = \sum_{n=0}^{m} (-1)^n (\pi, C^n(K;D)) \in \mathrm{GR}(D,\pi)$$

and rather obviously the cochain Euler characteristic belongs to the image of $\mathrm{Burn}(\pi) \longrightarrow \mathrm{GR}(D,\pi)$. It is definitely not true that each $(\pi, H^n(K;D))$ belongs to this image.

We mention next a natural homomorphism

$$W_*(D,\pi) \longrightarrow H^2(C_2;GR(D,\pi)) \ .$$

If (π,V,b) is an orthogonal representation of π over D then by definition

$$\mathrm{adj}_b\colon (\pi,V) \simeq (\pi,\mathrm{Hom}_D(V,D))$$

and V is D-projective. Thus $(\pi,V)^0 = (\pi,V)$ and thus we find $\mathrm{cl}(\pi,V) \in H^2(C_2;GR(D,\pi))$. If (π,V,b) is metabolic and $N \subset V$ is a π-invariant sub-module with $N = N^{\perp}$ then we receive a short exact sequence of $D(\pi)$-modules

$$0 \longrightarrow (\pi,N) \xrightarrow{\ i\ } (\pi,V) \xrightarrow{\ P\ } (\pi,\mathrm{Hom}_D(N,D)) \longrightarrow 0$$

where $P(v)(x) \equiv b(v,x)$. Thus in $GR(D,\pi)$, $(\pi,V) = (\pi,N) + (\pi,N)^0$. In this manner $(\pi,V,b) \longrightarrow \mathrm{cl}(\pi,V)$ yields a well-defined homomorphism $W_0(D,\pi) \longrightarrow H^2(C_2;GR(D,\pi))$. The skew-symmetric case is done by analogy. Now from the oriented bordism groups we had

$$w\colon \mathcal{O}_{2_*}(\pi) \longrightarrow W_*(Z,\pi) \ .$$

We shall next determine the image of the composition

$$\mathcal{O}_{2_*}(\pi) \longrightarrow W_*(Z,\pi) \longrightarrow H^2(C_2;GR(Z,\pi)) \ .$$

Suppose now that (M^{2m},π) is an action of π as a group of orientation preserving diffeomorphisms on a closed oriented 2m-manifold. We deal only with the case $D = Z$ as the field case is much simpler. We put

$$\chi(M,\pi) = \sum_{j=0}^{2m} (-1)^j (\pi,H^j(M;Z)) \in GR(Z,\pi) \ .$$

In addition $w[M,\pi] \in W_*(Z,\pi)$ is $(\pi,V) = (\pi,H^m(M;Z)/\mathrm{Tor}))$ where the cup-product, thanks to Poincare duality, defines the $(-1)^m$-symmetric, π-invariant innerproduct space structure. We prove

(1.2) Theorem: For (M^{2m},π)

$$\chi(M,\pi)^0 = \chi(M,\pi)$$

and

$$cl(\pi,V) = cl(\chi(M,\pi)) \in H^2(C_2;GR(Z,\pi)) ,$$

therefore the image of the composite homomorphism

$$\mathcal{O}_{2_*}(\pi) \longrightarrow W_*(Z,\pi) \longrightarrow H^2(C_2;GR(Z,\pi))$$

agrees with the image of

$$\mathrm{Burn}(\pi)/2\ \mathrm{Burn}(\pi) \longrightarrow H^2(C_2;GR(Z,\pi)) .$$

Proof. We let $(\pi,T^j(M)) \subset (\pi,H^j(M;Z))$ denote the Z-torsion submodule. Actually the full import of Poincare duality shows

$$(\pi,H^{2m-j}(M;Z)/T^{2m-j}(M)) \simeq (\pi,\mathrm{Hom}(H^j(M;Z),Z)$$
$$(\pi,T^{2m-j+1}(M)) \simeq (\pi,\mathrm{Hom}(T^j(M),Q/Z)) .$$

Therefore in $GR(Z,\pi)$ we may write

$$(\pi,H^{2m-j}(M;Z)) = (\pi,H^j(M;Z))^0 + (\pi,T^{2m-j}(M)) + (\pi,T^{2m-j+1}(M))$$

so that in particular

$$(\pi,H^m(M;Z)) = (\pi,H^m(M;Z))^0 + (\pi,T^m(M)) + (\pi,T^{m+1}(M)) .$$

Next we look at the relation

$$\begin{aligned}
\chi(M,\pi) &= \sum_{j=0}^{2m} (-1)^j (\pi, H^j(M;Z)) \\
&= \sum_{j=0}^{m-1} (-1)^j (\pi, H^j(M;Z)) + (-1)^m (\pi, H^m(M;Z)) \\
&+ \sum_{j=0}^{m-1} (-1)^j (\pi, H^{2m-j}(M;Z)) \\
&= \sum_{j=0}^{m-1} (-1)^j [(\pi, H^j(M;Z)) + (\pi, H^j(M;Z))^0] \\
&+ \sum_{j=0}^{m-1} (-1)^j [(\pi, T^{2m-j}(M)) + (\pi, T^{2m-j+1}(M))] \\
&+ (-1)^m (\pi, H^m(M;Z)) .
\end{aligned}$$

Since $T^{2m+1}(M) = 0$ it will follow from the obvious cancellations that

$$\begin{aligned}
\chi(M,\pi) &= \sum_{j=0}^{m-1} (-1)^j [(\pi, H^j(M;Z)) + (\pi, H^j(M;Z))^0] \\
&+ (-1)^m (\pi, H^m(M;Z)) + (-1)^{m+1} (\pi, T^{m+1}(M)) .
\end{aligned}$$

We now observe that

$$\begin{aligned}
&(-1)^m (\pi, H^m(M;Z)) + (-1)^{m+1} (\pi, T^{m+1}(M)) \\
&= (-1)^m (\pi, V) + (-1)^m [(\pi, T^m(M)) + (\pi, T^m(M))^0]
\end{aligned}$$

so that

$$\begin{aligned}
\chi(M,\pi) &= (-1)^m (\pi, V) + \sum_{j=0}^{m-1} (-1)^j [(\pi, H^j(M;Z)) + (\pi, H^j(M;Z))^0] \\
&+ (-1)^m [(\pi, T^m(M)) + (\pi, T^m(M))^0] .
\end{aligned}$$

It is now certain that $\chi(M,\pi)^0 = \chi(M,\pi)$ and $cl(\chi(M,\pi)) = cl(\pi, V) \in H^2(C_2; GR(Z,\pi))$.

It is taken that the action of π on M^{2m} can be made simplicial. Thus by our previous discussions $cl(\pi,V) = cl(M,\pi)$ lies in the image of $\mathrm{Burn}(\pi)/2\,\mathrm{Burn}(\pi) \longrightarrow H^2(C_2; GR(Z,\pi))$. We leave to the reader to show the image of

$$\mathcal{O}_{2_*}(\pi) \longrightarrow H^2(C_2;GR(Z,\pi))$$

agrees exactly with the image of the Burnside ring. □

Now by analogy with $W_*(Q/Z,\pi)$ there is $GR(Q/Z,\pi)$, the Grothendieck group of actions of π as a group of automorphisms on finite abelian groups. The duality involution on $GR(Q/Z,\pi)$ may be taken to be

$$(\pi,A)^0 \equiv -\mathrm{Ext}^1_Z(A,Z) = -(\pi,\mathrm{Hom}_Z(A,Q/Z)) .$$

Swan showed [SW2] that for π finite

$$GR(Q/Z,\pi) \longrightarrow GR(Z,\pi) \longrightarrow GR(Q,\pi) \longrightarrow 0$$

is exact (as a matter of fact as a sequence of C_2-modules). Furthermore, $GR(Q/Z,\pi) \simeq \bigoplus_p GR(F_p,\pi)$ where p runs over all rational primes. What actually happens here is $GR(Z(1/p)/Z,\pi) \simeq GR(F_p,\pi)$ and $GR(Q/Z,\pi) \simeq \bigoplus_p GR(Z(1/p)/Z,\pi)$.

Suppose now $C_{p^r} \subset \pi$ is a central cyclic subgroup. We shall define a homomorphism $\eta\colon GR(Z,\pi) \longrightarrow GR(Z(1/p)/Z,\pi/C_{p^r})$. If (π,M) is any $Z(\pi)$-module then $H^1(C_{p^r};M)$ and $H^2(C_{p^r};M)$ are both finite abelian p-groups with induced actions of π/C_{p^r}. We let

$$\eta(\pi,M) = (\pi/C_{p^r}, H^1(C_{p^r};M)) - (\pi/C_{p^r}; H^2(C_{p^r};M))$$

in $GR(Z(1/p)/Z,\pi/C_{p^r})$. If $0 \longrightarrow (\pi,M_1) \longrightarrow (\pi,M) \longrightarrow (\pi,M_2) \longrightarrow 0$ is a short exact sequence of $Z(\pi)$-modules then the resulting exact hexagon, supressing π/C_{p^r},

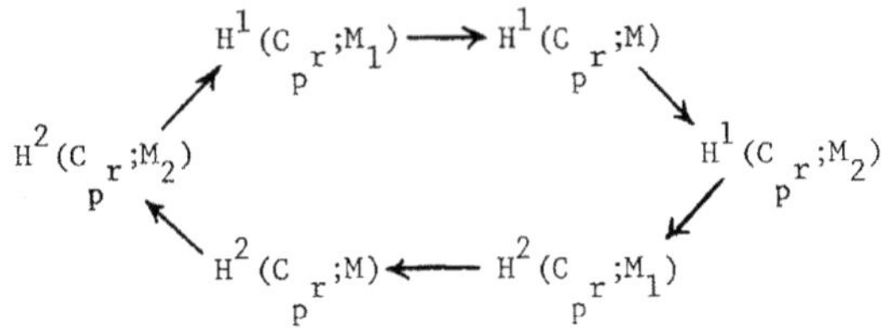

will show in $GR(Z(1/p)/Z,\pi/C_{p^r})$ that

$$H^1(C_{p^r};M) - H^2(C_{p^r};M)$$
$$= H^1(C_{p^r};M_1) - H^2(C_{p^r};M_1) + H^1(C_{p^r};M_2) - H^2(C_{p^r};M_2) .$$

Thus η is well defined.

We can also recall in the proof of IV.1.2 we showed in $GR(Z,\pi)$ that $(\pi,H^m(M;Z)/T^m(M)) = \Sigma\, (-1)^n(\pi,C^n(M))$. Hence in $GR(Z(1/p)/Z,\pi/C_{p^r})$ we should find

$$H^1(C_{p^r};H^m(M;Z)/T^m(M)) - H^2(C_{p^r};H^m(M;Z)/T^m(M))$$
$$= \sum_{n=0}^{2m} (-1)^n H^1(C_{p^r};C^n(M)) - \sum_{m=0}^{2m} (-1)^n H^2(C_{p^r};C^n(M)) .$$

Remember from Chapter II.1.2

$$(\pi/C_{p^r},H^i(C_{p^r};H^m(M)/T^m(M))$$

admits an invariant $Z(1/p)/Z$ valued innerproduct. Because of the definition of duality involution we receive in

$$H^1(C_2;GR(Z(1/p)/Z,\pi/C_{p^r}))$$

the difference

$$cl(H^1(C_{p^r};H^m(M;Z)/T^m(M))$$
$$- cl(H^2(C_{p^r};H^m(M;Z)/T^m(M)) .$$

But remember we also showed each $C^n(M)$ has a π-invariant Z-valued innerproduct. The upshot is

(1.3) Proposition: In $H^1(C_2;GR(Z(1/p)/Z,\pi/C_{p^r})$ there is the relation

$$cl(H^*(C_{p^r};H^m(M^{2m};Z)/TOR))$$
$$= cl(H^*(C_{p^r};C^*(M))) .$$

Here $H^*(C_{p^r};-) = H^1(C_{p^r};-) \oplus H^2(C_{p^r};-)$. We need not concern ourselves with signs as $H^1(C_2,GR(Z(1/p)/Z,\pi/C_{p^r}))$ is surely an elementary abelian 2-group. Later we shall explain how we plan to apply IV.1.3 before actually doing so. □

If $\pi = C_{p^k}$ a cyclic group of odd prime power order then Swan has determined $GR(Z,C_{p^k})$ so that $H^2(C_2;GR(Z,C_{p^k}))$ is computable. We shall summarize the result. For $0 \leq j \leq k$, let $\mathscr{C}(j)$ denote the ideal class group of the p^j-cyclotomic number field. Using complex conjugation an action of C_2 on $\mathscr{C}(j)$ is given by $A \to \bar{A}^{-1}$ where A is a fractional ideal in the number field. Then we can state

(1.4) Theorem: (Swan) There is an isomorphism

$$H^2(C_2;GR(Z,C_{p^k})) \simeq [\prod_{j=0}^{k} H^2(C_2;\mathscr{C}(j))] \oplus Burn(C_{p^k})/2\,Burn(C_{p^k}) .$$

The quotient $Burn(C_{p^k})/2\,Burn(C_{p^k})$ is an F_2 vector space of dimension $k+1$. A basis is provided by $(C_{p^k}/C_{p^j},C_{p^k})$ with $0 \leq j \leq k$. □

The group C_{29} exhibits the first anomaly in that $H^2(C_2;\mathscr{C}(1))$ is an elementary abelian 2 group of rank 3 (computation following Kummer).

Following Swan we may, in fact for D a Dedekind domain or a field, obtain $GR(D,\pi)$ by restricting ab initio to finitely generated $D(\pi)$-modules which are D-projective and to short exact sequences of such. In this format it is clear that $GR(D,\pi)$ is a commutative ring with identity and that the duality involution is an automorphism of this ring structure. Hence $W_*(D,\pi) \longrightarrow H^2(C_2;GR(D,\pi))$ is a ring homomorphism for any Dedekind domain or field.

2. The characteristic polynomial of an isometry

The purpose of this section is the discussion of the properties of the characteristic polynomial of an isometry and to compute $H^2(C_2;GR(F_p,C_\ell))$ where $(p,\ell)=1$. We let D denote a Dedekind domain or a field. We define a ring $\mathbf{P}(D)$ as follows. First $\mathscr{P}$ will denote all monic polynomials of positive degree whose constant term, a_0, is a unit in D^*. With respect to ordinary polynomial multiplication $\mathscr{P}$ is a commutative cancellation semi-group without identity. Even more it enjoys a unique factorization property. Any polynomial $q(t) \in \mathscr{P}$ can be uniquely written as a product of distinct factors

$$q(t) = (p_1(t))^{r_1} \dots (p_k(t))^{r_k}$$

with positive exponents and each $p_j(t) \in \mathscr{P}$ irreducible over the field of fractions of D. We allow $\mathscr{P}$ to generate a free abelian group and we factor out the subgroup generated by all relations of the form $q_1(t)q_2(t) - q_1(t) - q_2(t)$. In the quotient group $\mathbf{P}(D)$ addition corresponds to polynomial multiplication. Additively $\mathbf{P}(D)$ is the free abelian group generated by the irreducible polynomials in $\mathscr{P}(D)$.

The ring structure is discovered as follows. There is the algebraic closure $\mathscr{K}$ of the field of fractions of D. If $p_1(t), p_2(t) \in \mathscr{P}$ they factor in $\mathscr{K}$, allowing repetitions, as

$$p_1(t) = (t-\lambda_1) \dots (t-\lambda_n)$$
$$p_2(t) = (t-\mu_1) \dots (t-\mu_m) .$$

Then $(p_1 \circ p_2)(t)$ is the polynomial $\prod_{i,j}(t-\lambda_i\mu_j)$. Use the symmetric function twice to show this polynomial lies in $\mathbb{P}(D)$, and thus products are introduced into $\mathbb{P}(D)$. Actually we can embed $\mathbb{P}(D)$ into the integral group ring $Z(\mathcal{K}^*)$ by factoring

$$q(t) = (t-\lambda_1)^{m_1} \cdots (t-\lambda_k)^{m_k}$$

into distinct roots and then sending $q(t)$ into $m_1\lambda_1 + \ldots + m_k\lambda_k$. Thus $t-1 \in \mathbb{P}(D)$ plays the role of the multiplicative identity.

On $\mathbb{P}(D)$ we introduce the involution $p^*(t) = a_0^{-1}t^n p(t^{-1})$ wherein $n = \deg p(t)$. This agrees with the canonical involution on $Z(\mathcal{K}^*)$ induced by sending an element of $\mathcal{K}^*$ into its inverse. In this manner $H^2(C_2;\mathbb{P}(D))$ becomes a linear algebra over F_2.

As remarked earlier, $\mathbb{P}(D)$ is simply the free abelian group generated by the irreducible polynomials in $\mathcal{P}$. The action of C_2 permutes this basis, hence as an F_2 vector space there is a basis of $H^2(C_2;\mathbb{P}(D))$ in 1-1 correspondence with the irreducible polynomials in $\mathcal{P}$ that satisfy $p^*(t) = p(t)$.

The motivation for the involution on $\mathbb{P}(D)$, following [LdM], is this. If $L: V \to V$ is an automorphism of a finitely generated free D-module, then there are characteristic polynomials $p_L(t)$ and $p_{L^{-1}}(t)$ which belong to $\mathcal{P}$. We claim that $p_{L^{-1}}(t) = p_L^*(t)$. Note that

$$(-tL^{-1})(t^{-1}I-L) = tI-L^{-1}$$

so

$$p_{L^{-1}}(t) = \det(-tL^{-1})\det(t^{-1}I-L)$$

and

$$\det(-tL^{-1}) = (-1)^n t^n (-1)^n a_0^{-1} = a_0^{-1}t^n .$$

(2.1) <u>Lemma</u>: <u>If</u> $p(t) \in \mathcal{P}$ <u>then</u> $p^*(t) = p(t)$ <u>if and only if</u> $a_j = a_0 a_{n-j}$, $0 \le j \le n$. <u>In particular if</u> $p^*(t) = p(t)$ <u>then</u> $a_0^2 = a_n = 1$ <u>so</u> $a_0 = \pm 1$.

(2.2) Lemma: If $p^*(t) = p(t) \in \mathcal{P}$ and deg $p(t)$ is odd then $p(-a_0) = 0$. If deg$(p(t))$ is even, $a_0 = -1$ and $ch(D) \neq 2$, then $p(1) = 0$.

Proof. If $n = 2k+1$ then

$$p(-a_0) = \sum_{j=0}^{n} a_j(-a_0)^j = \sum_{j=0}^{k} (a_j(-a_0)^j + a_j a_0(-a_0)^{j+1})$$
$$= \sum_{j=0}^{k} a_j((-a_0)^j + a_0(-a_0)^{j+1}) = 0$$

since $a_0 = \pm 1$. When $n = 2k$, $a_0 = -1$ then $a_k = -a_k = 0$ and $p(1) = \sum_{j=0}^{k-1} a_j - a_j = 0$. □

(2.3) Corollary: If $p^*(t) = p(t) \in \mathcal{P}$ is irreducible then deg$(p(t))$ is even and $a_0 = 1$ unless $p(t) = t \pm 1$.

If $p(t)$ is irreducible the quotient $D[t]/(p(t)) = D(\theta)$ is an integral domain (or a field if D were a field).

(2.4) Lemma: If $p^*(t) = p(t) \in \mathcal{P}$ is irreducible then $D(\theta)$ inherits an involution for which $\hat{\theta} = \theta^{-1}$.

Proof. We introduce the ring of finite Laurent series $D[t,t^{-1}]$ which admits the involution $\hat{t} = t^{-1}$. If degree $p(t) = 2k$ let $\gamma(t) = t^{-k}p(t)$ so that $\hat{\gamma} = \gamma$ and $D[t,t^{-1}]/(\gamma(t)) \simeq D[t]/(p(t))$. The involution on $D[t,t^{-1}]$ is now passed onto $D(\theta)$. □

For $p(t) = t \pm 1$ we just receive the identity involution on D. Now consider $GR(F,C_n)$ where n is an (odd) integer. We first note $(t^n-1)^* = t^n - 1$. If $p(t)$ is irreducible and $p(t)|t^n-1$ then $p^*(t)$ is also an irreducible divisor of t^n-1. Each irreducible factor produces an irreducible representation of C_n on the extension field $F(\theta) = F[t]/(p(t))$ where the generator of C_n acts as multiplication by θ.

If $p^*(t) = p(t)$, and F is perfect, we can identify $(C_n, F(\theta))$ with $(C_n, \mathrm{Hom}_F(F(\theta),F))$. For $w \in F(\theta)$ define $\varphi_w(v) = tr_{F(\theta)/F}(v\hat{w})$, where $w \to \hat{w}$ is the involution induced by $\theta \to \theta^{-1}$. Then $\varphi_{Tw}(v) = tr(v\widehat{\theta w}) = tr(\theta^{-1}v\hat{w}) = \varphi_w(T^{-1}v)$ so we have the identification.

There is the homomorphism of $GR(F,C_n)$ into $\mathbb{P}(F)$ which to each representation assigns the characteristic polynomial of the generator T of C_n. This commutes with the involutions on $GR(F,C_n)$ and $\mathbb{P}(F)$. Furthermore there is a homomorphism of the free abelian group generated by irreducible factors of t^n-1 back into $GR(F,C_n)$. To each irreducible $p(t) \mid t^n-1$ we assign the irreducible representation $F(\theta)$. In fact the characteristic polynomial provides an isomorphism of $GR(F,C_n)$ with the subring of $\mathbb{P}(D)$ generated by the irreducible factors of t^n-1 in $F[t]$. Notice this always includes $t-1$, the multiplicative identity in $\mathbb{P}(D)$.

Suppose now that F has finite characteristic, p, and that $n = p^r \ell$ with $(p,\ell) = 1$ and $r > 0$. Then

$$(t^n-1) = ((t^\ell)^{p^r}-1) = (t^\ell-1)^{p^r}$$

in $F[t]$. This tells us that the irreducible factors of (t^n-1) are the same as those of $t^\ell-1$. This translates into a statement about Grothendieck rings as follows. Let $C_n \xrightarrow{\partial} \tilde{C}_\ell$ be the quotient homomorphism obtained by factoring out the $C_{p^r} \subset C_n$ generated by T^ℓ. This induces $\nu\colon GR(F,\tilde{C}_\ell) \longrightarrow GR(F,C_n)$.

(2.5) <u>Lemma</u>: *If* $ch(F) = p$ *and* $n = p^r\ell$, $r > 0$ *and* $(p,\ell) = 1$ *then* $\nu\colon GR(F,\tilde{C}_\ell) \simeq GR(F,C_n)$.

The effect of this result is that we may compute under the assumption that the field characteristic is relatively prime to the group order. For a general field we can also mention

(2.6) <u>Lemma</u>: *The assignment of the characteristic polynomial produces an isomorphic embedding*

$$H^2(C_2;GR(F,C_n)) \longrightarrow H^2(C_2;\mathbb{P}(F))$$

whose image is generated by the classes of irreducible divisors of t^n-1 *in* $F[t]$ *for which* $p^*(t) = p(t)$.

The image of the Burnside ring under the composition

$$\mathrm{Burn}(C_n) \longrightarrow GR(F, C_n) \longrightarrow \mathbb{P}(F)$$

is the subring generated by the polynomials t^m-1 where $m|n$.

For a rational prime p and $(\ell,p)=1$ the computation of $GR(F_p, C_\ell)$ is intimately related to the decomposition of p in the various cyclotomic number fields $Q(\lambda_d), d|\ell$ and $\lambda_d = \exp(2\pi i/d)$. Let us go into this. First in $Z[t]$ we have the factorization

$$t^\ell - 1 = \prod_{d|\ell} \Phi_d(t)$$

into the product of cyclotomic polynomials. For each $\Phi_d^*(t) = \Phi_d(t)$. Now when $\Phi_d(t)$ is reduced into $F_p[t]$ it may factor, however it is true that with $(p,\ell)=1$ the cyclotomic factors $\Phi_d(t)$ do remain pairwise relatively prime. This is seen by noting that $t^\ell-1$ and its derivative $\ell t^{\ell-1}$ have no common roots and, therefore, $t^\ell-1$ has no repeated roots. Thus we concentrate our attention on each $\Phi_d(t)$ in $F_p[t]$. There is a factorization

$$\Phi_d(t) = f_{1,d}(t) \ldots f_{r,d}(t)$$

into distinct irreducible polynomials. According to Kummer these are in 1-1 correspondence with the prime ideals $\mathcal{P}_1, \ldots, \mathcal{P}_r$ in $Z(\lambda_d)$ into which the rational prime p factors. That is,

$$(p) = \mathcal{P}_1 \ldots \mathcal{P}_r .$$

Since $(p,\ell)=1$ the ramification indices are all 1. In addition $\mathcal{P}_j$ is generated by p together with $\hat{f}_{j,d}(\lambda_d)$ where $\hat{f}_{j,d}(t)$ is any monic integral polynomial whose mod p reduction is $f_{j,d}(t)$. The residue field $Z(\lambda_d)/\mathcal{P}_j$ is $F_p(\theta) = F_p[t]/(f_{j,d}(t))$. The action of the generator T of C_ℓ on the residue field is multiplication by λ_d.

All of the factors $f_{j,d}$ have the same degree, f. It is the smallest positive integer for which $p^f \equiv 1 \pmod d$. The number of distinct factors, r, is calculated from $rf = \varphi(d)$, the degree of the cyclotomic extension.

The fundamental question then is when does $f^*_{j,d}(t) = f_{j,d}(t)$? Put otherwise, equivalently, when is $\bar{\mathscr{P}}_j = \mathscr{P}_j$? Since the abelian Galois group $G(Q(\lambda_d)/Q)$ transitively permutes the $\mathscr{P}_j$ we see that either all $f_{j,d}(t)$ satisfy $f^*_{j,d}(t) = f_{j,d}(t)$ or none do.

Remember p is unramified in $Q(\lambda_d)$. The decomposition group of $\mathscr{P}_j$ is generated by the automorphism $\lambda_d \longrightarrow \lambda_d^p$. This is exactly the Artin symbol $(p, Q(\lambda_d)/Q) \in G(Q(\lambda_d)/Q)$. Furthermore this decomposition group is mapped isomorphically onto the Galois group of the residue field over F_p. Thus we can have an involution of $F_p(\theta) = F_p[t]/(f_{j,d}(t))$ that satisfies $\hat{\theta} = \theta^{-1}$ if and only if complex conjugation, $\lambda_d \longrightarrow \lambda_d^{-1}$, lies in the decomposition group of $\mathscr{P}_j$.

(2.7) <u>Lemma</u>: <u>If in</u> $F_p[t]$, $\Phi_d(t) = \prod_{j=1}^{r} f_{j,d}(t)$ <u>then</u> $f^*_{j,d}(t) = f_{j,d}(t)$ <u>if and only if</u> $gr(p) \subset Z_d^*$, <u>the cyclic subgroup in</u> Z_d^* <u>generated by</u> p, <u>contains</u> -1.

There is another way to prove this lemma involving less number theory. The involution on the ring of Laurent polynomials $F_p[t,t^{-1}]$ taking t to t^{-1} induces a field isomorphism of $F_p(\theta) = F_p[t,t^{-1}]/(f_{j,d}(t)) \cong F_p[t,t^{-1}]/(f^*_{j,d}(t))$. If $f^*_{j,d}(t) = f_{j,d}(t)$, the fields are the same and we have an automorphism carrying the primitive d^{th} root of unity, θ, to its inverse. Since the Galois group $G(F_p(\theta)/F_p)$ is cyclic generated by the Frobenius automorphism $x \longrightarrow x^p$, it follows that -1 is equivalent to some power of p modulo d. On the other hand, if $-1 \in (p) \subset Z_d^*$, there is an automorphism of $F_p(\theta)$ taking θ to its inverse. It follows that θ must be a root of both $f_{j,d}$ and $f^*_{j,d}$ in $F(\theta)$, since the roots of $f^*_{j,d}$ are the inverses of the roots of $f_{j,d}$. Hence $f^*_{j,d}(t) = f_{j,d}(t)$.

This was used previously in computing the torsion subgroup of $\mathscr{H}_o(Z(\lambda_k/k))$. We can even refine the structure description of $GR(F_p, C_\ell)$. For each $d|\ell$ let $GR(F_p, \Phi_d)$ be the Grothendieck group of representations (C_ℓ, V) in which $\Phi_d(T) = 0$. Then

$$GR(F_p,C_\ell) = \sum_{d|\ell} GR(F_p,\Phi_d) .$$

In addition

$$H^2(C_2;GR(F_p,C_\ell)) \simeq \sum_{d|\ell} H^2(C_2;GR(F_p,\Phi_d))$$

however we only receive a contribution from those divisors at which $gr(p) \subset Z_d^*$ contains -1.

Now via $C_n \xrightarrow{\nu} \tilde{C}_d$ we have for each $d|n$ the representation $Z(\Lambda_d)/(p) = F_p(\Lambda_d) = \sum_{j=1}^{r} Z(\Lambda_d)/\mathcal{P}_j$. What we can say is

(2.8) Lemma: The $cl(F_p(\Lambda_d))$ in $H^2(C_2;GR(F_p,C_\ell))$ is non-zero if and only if -1 lies in $gr(p) \subset Z_d^*$.

These particular cohomology classes are quite significant in the following chapter.

For any finite field, F, we can calculate $W_*(F,C_n)$ for a cyclic group in terms of $H^2(C_2;GR(F,C_n))$. To compute we know from chapter I that all we need do is identify the maximal ideals $\mathcal{M} \subset F[C_n]$ for which $\hat{\mathcal{M}} = \mathcal{M}$. These are precisely the principal ideals $(q(T))$ where $q(t)$ is an irreducible factor of t^n-1 in $F[t]$ which satisfies $q^*(t) = q(t)$. These will include $(T-I)$ which yields us a copy of $W_*(F)$. Otherwise $\deg q(t) = 2k$ so we write $\gamma(T) = T^{-k}q(T)$ so $\hat{\gamma} = \gamma$ and $\mathcal{M}/(\gamma) = F(\theta)$ with a non-trivial involution. Thus

(2.9) Theorem: For a finite field

$$W_0(F,C_n) \simeq W(F) \oplus \sum \mathcal{H}_o(F(\theta))$$

$$W_2(F,C_n) \simeq \sum \mathcal{H}_2(F(\theta)), \quad ch(F) \neq 2 .$$

If $ch(F) = 2$ then $W_0(F,C_n) = W_2(F,C_n)$. The only invariant for $\mathcal{H}_o(F(\theta))$ is rank mod 2 over $F(\theta)$. We have the homomorphism

$$W_0(F,C_n) \longrightarrow H^2(C_2;GR(F,C_n)) \longrightarrow H^2(C_2;\mathbb{P}(F)) .$$

For $[1] \in \mathcal{H}_0(F(\theta))$ the characteristic polynomial of T, that is, of multiplication by θ, is $q(t)$, so $[1] \longrightarrow cl(q(t)) \in H^2(C_2;\mathbb{P}(F))$.

(2.10) *Corollary: For a finite field of* $ch(F) \neq 2$

$$0 \longrightarrow Z/2Z \longrightarrow W_0(F,C_n) \longrightarrow H^2(C_2;GR(F,C_n)) \longrightarrow 0$$
$$0 \longrightarrow W_2(F,C_n) \longrightarrow H^2(C_2;GR(F,C_n)) \longrightarrow Z/2Z \longrightarrow 0$$

while if $ch(F) = 2$

$$W_0(F,C_n) \simeq W_2(F,C_n) \simeq H^2(C_2;GR(F,C_n)) .$$

When $ch(F) \neq 2$ *then the kernel of* $W_0(F,C_n) = W(F) \oplus \sum \mathcal{H}_0(F(\theta)) \longrightarrow H^2(C_2;GR(F,C_n))$ *is the fundamental ideal in* $W(F)$.

We might also make this observation. Suppose $n = p^r\ell$, $r > 0$, $(p,\ell) = 1$ and $m = p^jk$, $k|\ell$, $0 \leq j \leq r$. Then there is a commutative diagram of quotient homomorphisms

$$\begin{array}{ccc} C_n & \longrightarrow & \tilde{C}_m \\ \downarrow & & \downarrow \\ \tilde{C}_\ell & \longrightarrow & \tilde{C}_k \end{array} .$$

(2.11) *Lemma: In this situation there is a commutative cube*

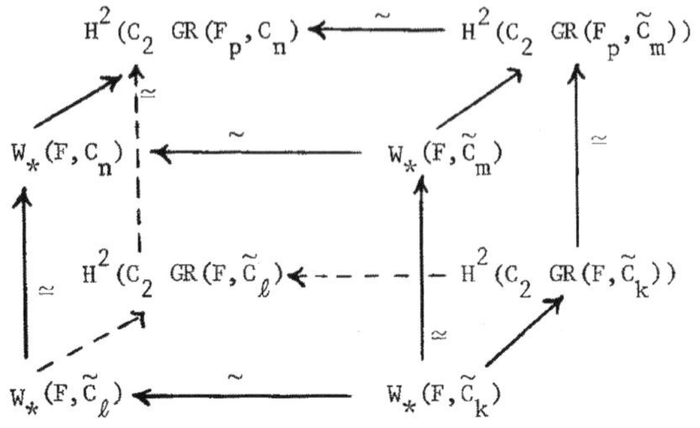

That the indicated monomorphisms do in fact have trivial kernels is elementary. To see $\nu\colon C_n \longrightarrow \tilde{C}_\ell$ induces an isomorphism

$$\nu\colon W_*(F,\tilde{C}_\ell) \simeq W_*(F,C_n)$$

the reader may show that if (C_n,V,b) is anisotropic over F, $ch(F)=p$, then the subgroup C_{p^r} acts trivially. (Hint: A representation of a p-group over a field of characteristic p always has a non-zero fixed vector.)

3. Invariants detecting torsion in $W_*(Q,C_n)$

We deal only with the symmetric case. The skew symmetric case is analogous and is left for the reader. Our viewpoint begins by introducing an isomorphism

$$\mathrm{Tr}\colon \sum_{m|n} \mathcal{H}_o(Q(\lambda_m)) \simeq W_0(Q,C_n)$$

which is a bit different from the one used in chapter III and is more convenient here. Specifically if $(V,[\ ,\ ])$ is a Hermitian innerproduct space over $Q(\lambda_m)$ then the rational form on V is given by $(x,y) = \frac{1}{m}\,\mathrm{tr}_{Q(\lambda_m)/Q}[x,y]$ and the generator T of C_n acts as multiplication by λ_m. Note explicitly the factor $1/m$. The effort in this section is devoted to the following problem. If $V \in \mathcal{H}_o(Q(\lambda_m))$ then there is $\bar{\partial}(V) \in \mathcal{H}_o(Q(\lambda_m)/Z(\lambda_m))$ and $\partial(V) = \partial(\mathrm{Tr}(V)) \in W_0(Q/Z,C_n)$. How can we compare these two invariants? The answer is through the cohomology group of C_2

acting on the Grothendieck group $GR(Q/Z,C_n)$. Certainly $W_0(Q/Z,C_n) \simeq \bigoplus_q W_0(F_q,C_n)$ and in (IV.2.10) we discussed the homomorphism

$$W_0(F_q,C_n) \longrightarrow H^2(C_2;GR(F_q,C_n)) .$$

Thus we should focus attention on $\mathcal{H}(Q(\Lambda_m)/Z(\Lambda_m))$. This Hermitian Witt group splits up as follows. Let $\mathfrak{Q} \subset Z(\Lambda_m)$ be a prime ideal for which $\overline{\mathfrak{Q}} = \mathfrak{Q}$, then the residue field $F(\mathfrak{Q}) = Z(\Lambda_m)/\mathfrak{Q}$ receives an involution and

$$\mathcal{H}_o(Q(\Lambda_m)/Z(\Lambda_m)) \simeq \bigoplus_{\mathfrak{Q}=\overline{\mathfrak{Q}}} \mathcal{H}_o(F(\mathfrak{Q})) .$$

The prime ideal $\mathfrak{Q}$ lives over a rational prime q. We claim there is

$$\mathcal{H}_o(F(\mathfrak{Q})) \longrightarrow H^2(C_2;GR(F_q,C_n)) .$$

First of all there is a homomorphism $\mathcal{H}_o(F(\mathfrak{Q})) \rightarrow W_0(F_q,C_n)$. That is, if $(W,[\ ,\ ])$ is a Hermitian innerproduct space over $F(\mathfrak{Q})$ then the F_q-valued innerproduct on W is

$$(w,w_1) = tr_{F(\mathfrak{Q})/F_q}[w,w_1] .$$

If C_n acts on W as multiplication by λ_n then C_n is a group of isometries with respect to the F_q-innerproduct. Using the composition with

$$W_0(F_q,C_n) \longrightarrow H^2(C_2;GR(F_q,C_n))$$

we arrive at

$$\mathcal{H}_o(F(\mathfrak{Q})) \longrightarrow H^2(C_2;GR(F_q,C_n)) .$$

Suppose first that the induced involution on the residue field is non-trivial.

Then $\mathcal{H}_o(F(\mathfrak{Q})) \simeq Z/2Z$ and rank mod 2 is the only invariant. We need to say what happens to $[1] \in \mathcal{H}_o(F(\mathfrak{Q}))$. In other words, what is the characteristic polynomial of T acting on $F(\mathfrak{Q})$ as multiplication by λ_m? Simply recall that in $F_q[t]$ the cyclotomic polynomial $\Phi_m(t) = f_1(t)^e \ldots f_r(t)^e$ with $f_j^*(t) = f_j(t)$. Then a unique one of these irreducible factors, determined by $\mathfrak{Q}$, is the characteristic polynomial. In fact we can say

(3.1) Lemma: *If $\mathfrak{Q}_1, \ldots, \mathfrak{Q}_r$ are the prime factors of q in $Z(\lambda_m)$ and if $\overline{\mathfrak{Q}_j} = \mathfrak{Q}_j$ for $1 \leq j \leq r$ induces a non-trivial involution on each residue field then*

$$0 \longrightarrow \Sigma\,\mathcal{H}(F(\mathfrak{Q}_j)) \longrightarrow H^2(C_2; GR(F_q, C_n)) .$$

If $q \nmid n$ the image is the summand $H^2(C_2; GR(F_q, \Phi_m))$. If $q|n$ write $n = q^r \ell$, $r > 0$, $(q,\ell) = 1$ and $m = q^s k$, $s \geq 0$, $(q,k) = 1$ then the image in $H^2(C_2; GR(F_q, \tilde{C}_\ell))$ $\simeq H^2(C_2; GR(F_q, C_n))$ is the summand $H^2(C_2; GR(F_q, \Phi_k))$.

Note $k|\ell$ and $k \neq 1$ if $s > 0$ because we do not permit $m = q^s$ here. The rationale is the diagram of quotient homomorphisms:

$$\begin{array}{ccc} C_n & \longrightarrow & \tilde{C}_m \\ \downarrow & & \downarrow \\ C_\ell & \longrightarrow & \tilde{C}_k \end{array}$$

which induces, by IV.2.11

$$\begin{array}{ccc} H^2(C_2; GR(F_q, C_n)) & \xleftarrow{\ \sim\ } & H^2(C_2; GR(F_q, \tilde{C}_m) \\ \uparrow \simeq & & \uparrow \simeq \\ H^2(C_2; GR(F_q, \tilde{C}_\ell) & \xleftarrow{\ \sim\ } & H^2(C_2; GR(F_q, \tilde{C}_k) \end{array} .$$

At each $\mathfrak{Q}_j$ the residue field of $Q(\lambda_m)$ is exactly the same as the residue field of $Q(\lambda_k)$ at $\mathfrak{Q}_j \cap Z(\lambda_k)$. Thus replacing C_n by $\tilde{C}_k$ we use the first part of the lemma to establish the second. The one potentially bad case not covered by the lemma is $m = q^s$, $s > 0$. However this evaporates if we recall from chapter III we showed $\bar{\partial}: \mathcal{H}_o(Q(\Lambda_{q^s})) \to \mathcal{H}_o(Q(\Lambda_{q^s})/Z(\Lambda_{q^s})$ is trivial over the ramified prime, q.

Thus we can compare $\bar{\partial}(V)$ with $\partial \circ Tr(V)$ in $H^2(C_2, GR(Q/Z, C_n)) \simeq \sum_q H^2(C_2; GR(F_q, C_n))$. One caveat however is that while $\bar{\partial}(V)$ is uniquely determined in $H^2(C_2; GR(Q/Z, C_n))$ we are only dealing with the image of $\partial \circ Tr(V)$ under

$$W_0(Q/Z, C_n) \longrightarrow H^2(C_2; GR(Q/Z, C_n)) .$$

Now in $Q(\Lambda_n)$ define $J(n) \subset Q(\Lambda_n)$ to be the fractional ideal of all $x \in Q(\Lambda_n)$ for which

$$tr_{Q(\Lambda_n)/Q}(xZ(\Lambda_n)) \subset nZ .$$

If $D^{-1} \subset Q(\Lambda_n)$ is the inverse different then $nD^{-1} = J(n)$.

(3.2) Lemma: The ideal $J(n)$ is a principal integral ideal generated by

$$u_n = \prod_{p|n} (\Lambda_p - 1) ,$$

product over all primes dividing n.

Proof. If we write $n = p_1^{r_1} \ldots p_k^{r_k}$ then $Q(\Lambda_n)$ is the composition of linearly disjoint fields $Q(\Lambda_{p_j^{r_j}})$ which have pairwise relatively prime discriminants. Thus, as ideals in $Z(\Lambda_n)$ the product of the differents of the various $Q(\Lambda_{p_j^{r_j}})$ is the different D of $Q(\Lambda_n)$. It follows that $J(n) = J(p_1^{r_1}) \ldots J(p_k^{r_k})$.

Next to compute $J(p^r)$ we write

$$t^{p^r} - 1 = (t^{p^{r-1}} - 1)\Phi_{p^r}(t)$$

so that differentiating and making the substitution $t = \lambda_{p^r}$ yields

$$p^r \lambda_{p^r}^{p^r - 1} = (\lambda_{p^r})^{p^{r-1}} \Phi'_{p^r}(\lambda_{p^r}) .$$

Then $\lambda_p = (\lambda_{p^r})^{p^{r-1}}$ and $\Phi'_{p^r}(\lambda_{p^r})$ generates the different. Thus $p^r D^{-1} = (\lambda_p - 1)$. The lemma follows. □

(3.3) Lemma: Let P be any finitely generated projective $Z(\lambda_n)$-module. If $x \in Z(\lambda_n)$, $x \neq 0$, then

$$P/xP \simeq (Z(\lambda_n)/(x))^a$$

where $a = \dim_{Q(\lambda_n)} P \otimes_{Z(\lambda_n)} Q(\lambda_n)$.

Proof. Obviously this is true for a free $Z(\lambda_n)$-module. But then by the Steinitz structure theorem for projective $Z(\lambda_n)$-modules it would be enough to verify (3.2) for fractional ideals. By the strong approximation theorem [O'M] every ideal class contains a representative which is relatively prime to (x). The remainder of the argument is elementary. □

We continue now with our comparison problem. Let $P \subset V$ be a $Z(\lambda_n)$-lattice for the Hermitian innerproduct structure on V. Then by IV.3.2, $v_n P$ is a Z-lattice for the associated rational form on V. If $P^{\#}$ is the $Z(\lambda_n)$-dual of P and $(u_n P)^+$ is the Z-dual of $u_n P$ then

(3.4) Lemma:

$$P^{\#} = (u_n P)^+ .$$

Proof. The containment $P^{\#} \subset (u_n P)^+$ is obvious. Now from chapter III there is

$$P^{\#} \simeq \mathrm{Hom}_{Z(\lambda_n)}(P, Z(\lambda_n)) \simeq \mathrm{Hom}_Z(P, Z)$$

in which the composition sends $w \in P^{\#}$ into $v \to tr(\frac{u_n}{n}[v,w]) = \frac{1}{n} tr\,[u_n v,w] = (u_n v,w)$. Thus $(u_n P)^+ \subset P^{\#}$. □

(3.5) Lemma: For $V \in \mathcal{H}_o(Q(\lambda_n))$, $cl(\partial(V)) = cl(\bar{\partial}(V)) + a\; cl(Z(\lambda_n)/(u_n))$ in $H^2(C_2;GR(Q/Z,C_n))$ where $a = \dim_{Q(\lambda_n)} V$.

Proof. We have

$$0 \longrightarrow P/u_n P \longrightarrow P^{\#}/v_n P \longrightarrow P^{\#}/P \longrightarrow 0$$

we apply IV.3.3 to $P/u_n P$ and note $P^{\#}/P$ represents $\bar{\partial}(V)$ while $P^{\#}/u_n P$ represents $\partial(V)$. □

It will be necessary for us to know $cl(Z(\lambda_n)/(u_n))$ rather explicitly. In fact we shall begin with a general remark.

(3.6) Lemma: If $n = p^r \ell$ with $r > 0$ and $(\ell,p) = 1$ then

$$cl(Z(\lambda_n)/(1-\lambda_p)) = cl\; F_p(\lambda_\ell)$$

in $H^2(C_2;GR(F_p,C_n))$.

Proof. View $Z(\lambda_n)$ as the extension of $Z(\lambda_\ell)$ obtained by adjoining λ_{p^r}. Then $(\lambda_{p^r})^{p^{r-1}} = \lambda_p$ and we have the principal ideal $(1-\lambda_p) \subset Z(\lambda_n)$. We first assert $(1-\lambda_p) \cap Z(\lambda_\ell) = pZ(\lambda_\ell)$. This is seen by noting that the ideal $(1-\lambda_p)$ is invariant under the entire Galois group $G(Q(\lambda_n)/Q)$. That is, if $(j,n) = 1$ then $\frac{1-\lambda_p^j}{1-\lambda_p}$ is a unit. Thus $(1-\lambda_p) \cap Z(\lambda_\ell)$ is also invariant under $G(Q(\lambda_\ell)/Q)$. Clearly $(1-\lambda_p) \cap Z(\lambda_\ell)$ contains p and since p is unramified in $Q(\lambda_\ell)$ this can only mean $pZ(\lambda_\ell) = (1-\lambda_p) \cap Z(\lambda_\ell)$.

We next assert that the quotient $Z(\lambda_n)/(1-\lambda_p)$ is isomorphic to $F_p(\lambda_\ell)[C_{p^{r-1}}]$. We think of this as follows. Consider the group ring $Z(\lambda_\ell)[C_{p^r}]$.

Put $\Delta = I - \tau^{p^{r-1}}$ where τ generates C_{p^r}. If $C_{p^r} \subset C_n$ then $\tau = T^{\ell}$. Clearly

$$Z(\Lambda_\ell)[C_{p^r}]/(p,\Delta) \simeq F_p(\Lambda_\ell)[C_{p^{r-1}}] .$$

However in view of lemma (1.1) from chapter II we know that (p,Δ) contains $\Sigma = I + \tau^{p^{r-1}} + \ldots + (\tau^{p^{r-1}})^{p-1}$ so that we have a factoring

$$\begin{array}{ccc} Z(\Lambda_\ell)[C_{p^r}] & \longrightarrow & F_p(\Lambda_\ell)[C_{p^{r-1}}] \\ \downarrow & \nearrow & \\ Z(\Lambda_\ell)[C_{p^r}]/(\Sigma) = Z(\Lambda_n) & & \end{array}$$

and Δ in $Z(\Lambda_n)$ is $1-\lambda_p$.

If the quotient homomorphisms $C_n \longrightarrow \tilde{C}_\ell$ and $C_n \longrightarrow \tilde{C}_{p^{r-1}}$ are used to make $F_p(\Lambda_\ell)$ and $F_p[C_{p^{r-1}}]$ into C_n-modules then $F_p(\Lambda_\ell)[C_{p^{r-1}}]$ is $F_p(\Lambda_\ell) \otimes_{F_p} F_p[C_{p^{r-1}}]$. Thus the characteristic polynomial in $\mathbb{P}(F_p)$ associated to $F(\Lambda_\ell)[C_{p^{r-1}}]$ is $\Phi_\ell(t) \circ (t^{p^{r-1}} - 1) = \Phi_\ell \circ (t-1)^{p^{r-1}} = \Phi_\ell(t)$ since $(t-1)$ is the identity in $\mathbb{P}(F)$. This establishes the lemma. $\square$

(3.7) <u>Exercise</u>: <u>If</u> $1 \leq s \leq r$ <u>then show</u> $cl(Z(\Lambda_n)/(1-\lambda_{p^s})) = cl\, F_p(\Lambda_\ell)$.

Now we can find $cl(Z(\Lambda_n)/(u_n))$. Write $n = p_1^{r_1} \ldots p_k^{r_k}$ so the ideals $(1-\lambda_{p_j})$ are pairwise relatively prime. Hence by the Chinese Remainder Theorem

$$Z(\Lambda_n)/(u_n) \simeq \Sigma\, Z(\Lambda_n)/(1-\lambda_{p_j}) .$$

Write $\ell_j \cdot p_j^{r_j} = n$ with $(\ell_j, p_j) = 1$.

(3.8) <u>Lemma</u>: <u>Under the isomorphism</u>

$$H^2(C_2; GR(Q/Z, C_n)) \simeq \bigoplus_q H^2(C_2; GR(F_q, C_n))$$

<u>the</u> $cl(Z(\Lambda_n)/(u_n))$ <u>is equal to</u> $\Sigma\, cl(F_{p_j}(\Lambda_{\ell_j}))$.

This will allow us to localize at each rational prime q. Of course at each q there will be

$$\partial_q : W_0(Q,C_n) \longrightarrow W_0(F_q,C_n) .$$

For $\bar{\partial}$, however, we should recognize that we are in the semi-local situation. If $\mathfrak{Q}_1,\ldots,\mathfrak{Q}_r$ are the distinct primes over q we would get, with $\mathfrak{Q}_j = \overline{\mathfrak{Q}}_j$

$$\bar{\partial}_q : \mathcal{H}(Q(\lambda_n)) \longrightarrow \Sigma\, \mathcal{H}(F(\mathfrak{Q}_j)) .$$

We may discard rational primes where $\mathfrak{Q}_j \neq \overline{\mathfrak{Q}}_j$.

The localization is carried out by replacing P with $P \otimes Z_{(q)}$. Or, if the reader prefers more formality, replace Z by $Z_{(q)}$ and replace $Z(\lambda_m)$ with $Z_{(q)}(\lambda_m)$, the integral closure of $Z_{(q)}$ in $Q(\lambda_m)$. This latter is a principal ideal domain (why?). Now use lattices over $Z_{(q)}(\lambda_n)$. The upshot is

(3.9) Lemma: *If* $n = p^r \ell$, $r > 0$, $(p,\ell) = 1$ *then* $\mathrm{cl}\,\partial_p(V) = \mathrm{cl}(\bar{\partial}_p(V)) + a\,\mathrm{cl}(F_p(\lambda_\ell))$ *while if* q *does not divide* n

$$\mathrm{cl}(\partial_q(V)) = \mathrm{cl}(\bar{\partial}_q(V)) .$$

All equations are taken in $H^2(C_2; GR(F_q,C_n))$.

Thus we know how to compare $\bar{\partial}$ and ∂. The problem now is this. It is true $\bar{\partial}$ is an invariant for $W_0(Q,C_n)$, but to define it we first have to back an element up into $\sum_{m|n} \mathcal{H}(Q(\lambda_m))$. Thus we ask if there are invariants defined directly on a rational (C_n,U,b) that would determine the $\bar{\partial}_q$. In view of IV.3.9 we only need to worry about $p|n$ since otherwise ∂_q determines $\bar{\partial}_q$.

(3.10) Definition: *If* (C_n,U,b) *is a rational orthogonal representation of* C_n *then a special* $Z_{(p)}$ *lattice* $L \subset U$ *is a* C_n *invariant* $Z_{(p)}$*-lattice for which*

a) L^+/L *is an* F_p*-vector space*

b) <u>the action of</u> $C_{p^r} \subset C_n$ <u>on</u> L^+/L <u>is trivial</u>.

Of course $n = p^r \ell$ with $r > 0$ and $(p,\ell) = 1$. Actually if $p \mid n$ we could just require condition a).

(3.11) <u>Lemma</u>: <u>Given any</u> Z-<u>lattice</u> $K \subset U$ <u>there is a special</u> $Z_{(p)}$-<u>lattice</u> $L \supset K$.

Proof. Start with an invariant Z-lattice K. Let $K_{(p)} = K \cdot Z_{(p)} = K \otimes_Z Z_{(p)}$. Then

$$K^+/K \otimes K_{(p)} = K^+_{(p)}/K_{(p)} = \overline{K}$$

is a finite abelian p-group which supports a C_n-invariant symmetric innerproduct with values in $Z_{(p)}(\frac{1}{p})/Z_{(p)}$.

Now take $H \subset \overline{K}$ to be an invariant subgroup which is maximal with respect to $H \subset H^\perp$. Then $(C_n, H^\perp/H)$ is anisotropic in the sense of chapter I. Therefore $H^\perp/H$ is an F_p-vector space and the subgroup C_{p^r} acts trivially. If we take

$$L = \{x \mid x \in K^+_{(p)},\ x + K_{(p)} \in H\}$$

then L is a special $Z_{(p)}$-lattice. □

Choose $t_s = T^{n/p^s}$ where $n = p^r \ell$ with $(p,\ell) = 1$ and $1 \leq s \leq r$. Thus t_s has order p^s. Put

$$\Delta_s = t_s^{\frac{p^s-1}{2}} - t_s^{\frac{p^s+1}{2}}$$

$$\Sigma_s = I + t_s + \ldots + t_s^{p^s-1}.$$

There is then θ_s with $\hat{\theta}_s = -\theta_s$ and $\Delta_s \theta_s = p^s - \Sigma_s$.

(3.12) <u>Lemma</u>: <u>If</u> $L \subset U$ <u>is a special</u> $Z_{(p)}$-<u>lattice then on</u>

$$\text{im}: H^1(C_{p^s};L) \longrightarrow H^1(C_{p^s};L^+)$$

<u>there is an innerproduct with values in</u> $Z_{(p)}(1/p)/Z_{(p)}$.

Proof. This finite innerproduct will in fact be skew-symmetric. From one viewpoint the invariant non-singular pairing $L \times L^+ \longrightarrow Z_{(p)}$ will induce, as in chapter II, the finite non-singular pairing

$$H^1(C_{p^s};L) \times H^1(C_{p^s};L^+) \longrightarrow Z_{(p)}(1/p)/Z_{(p)}$$

and this will produce the required innerproduct on image: $H^1(C_{p^s};L) \longrightarrow H^1(C_{p^s};L^+)$. There is an alternative approach to which we shall also appeal. For an endomorphism $\varphi: A \longrightarrow A$ we use A^{φ} to denote the kernel and $_{\varphi}A$ to denote the image. Since C_{p^s} acts trivially on L^+/L it follows that $_{\Delta_s}L^+ \subset L^{\Sigma_s}$. Clearly $L^{\Sigma_s}/_{\Delta_s}L^+$ can be identified with the im: $H^1(C_{p^s};L) \longrightarrow H^1(C_{p^s};L^+)$. Then on $L^{\Sigma_s}/_{\Delta_s}L^+$ we put

$$(\tilde{x},\tilde{y}) = \frac{1}{p^s}\, b(\theta x, y) \in Z_{(p)}(1/p)/Z_{(p)}$$

for $\tilde{x},\tilde{y} \in L^{\Sigma_s}/_{\Delta_s}L^+$. It is routine to verify this is a finite innerproduct. We can think of C_n acting on $L^{\Sigma_s}/_{\Delta_s}L^+$ to produce an element in $H^2(C_2;GR(F_p,C_n)) \simeq H^2(C_2;GR(F_p,C_\ell))$. □

In this manner we actually obtain for each prime dividing n a family of homomorphisms

$$\alpha(p^s,-): W_0(Q,C_n) \longrightarrow H^2(C_2;GR(F_p,C_n))$$

where $1 \leq s \leq$ multiplicity of p in n. These, together with all $\partial_q(U)$ will determine all $\bar{\partial}_q(U)$.

(3.13) <u>Lemma</u>: <u>Each</u> $\alpha(p^s,-)$ <u>is a well defined homomorphism of</u>

$W_0(Q,C_n) \longrightarrow H^2(C_2;GR(F_p,C_n)) \simeq H^2(C_2;GR(F_p,C_\ell))$.

Proof. Consider the case of $L \subset M \subset M^+ \subset L^+$ where L,M are special $Z_{(p)}$ lattices for (C_n,U,b). Then

$$_{\Delta_s}M^+ \subset {}_{\Delta_s}L^+ \subset L^{\Sigma_s} \subset M^{\Sigma_s} .$$

Work with the finite forms defined on $L^{\Sigma_s}/_{\Delta_s}L^+$ and $M^{\Sigma_s}/_{\Delta_s}M^+$. Suppose $x \in M^{\Sigma_s}/_{\Delta_s}M^+$ annihilates $L^{\Sigma_s}/_{\Delta_s}M^+$, then $x \in {}_{\Delta_s}L^+/_{\Delta_s}M^+$ and hence

$$({}_{\Delta_s}L^+/_{\Delta_s}M^+)^\perp = L^{\Sigma_s}/_{\Delta_s}M^+$$

and

$$cl(M^{\Sigma_s}/_{\Delta_s}M^+) = cl(({}_{\Delta_s}L^+/_{\Delta_s}M^+)^\perp/({}_{\Delta_s}L^+/_{\Delta_s}M^+)) = cl(L^{\Sigma_s}/_{\Delta_s}L^+) .$$

If L and M are just any two special $Z_{(p)}$-lattices in U we pass to U $-U$ and the special lattice $L \perp -M$. Since $U \perp -U$ is metabolic $L \perp -M \subset K$ where K is a special lattice with $K = K^+$ and the form $H^1(C_{p^s};K^+)$ is Witt trivial. Now $\alpha(p^s,-)$ is additive and thus $\alpha(p^s,L) - \alpha(p^s,M) = 0$. Consequently α is well defined. ☐

Next we are going to see how α-invariants relate to $\bar{\partial}_p(V)$ where V is a Hermitian innerproduct space over $Q(\lambda_{p^{j_k}})$, $k|\ell$ and $1 \leq j \leq r$. Remember V defines an element of $W_0(Q,C_n)$ where $Tv = \lambda_{p^{j_k}} \cdot v$ and $(v,w) = \frac{1}{p^{j_k}}\, tr_{Q(\lambda_{p^{j_k}})/Q}[v,w]$. Using IV.3.9

$$cl\ \partial_p(V) = cl(\bar{\partial}_p(V)) + a\ cl(F_p(\lambda_k))$$

with $a = \dim_{Q(\lambda_{p^{j_k}})} V$.

Let us suppose that $L \subset V$ is a special $Z_{(p)}$-lattice for the rational form on

V. Let $E = L^+/L$ and remember $cl(E) = cl(\partial_p(V))$. There is then the exact hexagon

$$\begin{array}{ccccc} & H^1(C_{p^s};L) & \longrightarrow & H^1(C_{p^s};L^+) & \\ \nearrow & & & & \searrow \\ H^2(C_{p^s};E) & & & & H^1(C_{p^s};E) \\ \nwarrow & & & & \swarrow \\ & H^2(C_{p^s};L^+) & \longleftarrow & H^2(C_{p^s};L) & \end{array}$$

By choice of L as special $H^1(C_{p^s};E) = E = H^2(C_{p^s};E)$. Because L and L^+ are $Z_{(p)}(\Lambda_{p^j k})$-projectives we see easily that

$$H^1(C_{p^s};L) = 0 = H^1(C_{p^s};L^+) \quad \text{if} \quad j \leq r-s$$
$$H^2(C_{p^s};L) = 0 = H^2(C_{p^s};L^+) \quad \text{if} \quad j > r-s\ .$$

If $j \leq r-s$ then C_{p^s} will act trivially on $L^+ \supset L$. If $j > r-s$ then nothing in $L \subset L^+$ can be fixed under C_{p^s}. Hence

$$\alpha(_{p^s};V) = \begin{cases} 0 & \text{if} \quad j \leq r\text{-}s \\ cl(E) + cl(H^1(C_{p^s};L^+)) & \text{if} \quad j > r\text{-}s. \end{cases}$$

If $a = \dim_{Q(\Lambda_{p^j}k)} V$ and $j > r\text{-}s$ then

$$\begin{aligned} cl\ H^1(C_{p^s};L^+) &= cl(L^+/(1\ \lambda_{p^{j+r-s}})L^+) \\ &= a\ cl(Z(\Lambda_{p^j k}))/(1-\lambda_{p^{j+s-r}}) \\ &= a\ cl(F_p(\Lambda_k))\ . \end{aligned}$$

For this last computation we use IV.3.3 and IV.3.7. Hence

(3.14) Lemma: If $j \leq r\text{-}s$ then $\alpha(p^s,V) = 0$ while if $j > r\text{-}s$

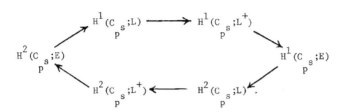

$$\alpha(p^s, V) = cl(E) + a(cl\ F_p(\Lambda_k)) = \bar{\partial}_p(V)\ .$$

(3.15) *Theorem*: *An element in* $W_0(Q, C_n)$ *is uniquely determined by*

1) *its multisignature*
2) *its image under* $\partial\colon W_0(Q, C_n) \longrightarrow W_0(Q/Z, C_n)$
3) *its* α*-invariants*.

Proof. First recall multisignature is a homomorphism

$$Sgn\colon W_0(Q, C_n) \longrightarrow Z(C_n)\ .$$

We form $U \otimes_Q C$ to obtain a representation of C_n on a Hermitian innerproduct space over C. Using the action of T this splits up orthogonally into eigenspaces. Each eigenspace, as a Hermitian innerproduct subspace, has a signature.

Now $Tr\colon \sum_{m|n} \mathcal{H}_o(Q(\Lambda_m)) \simeq W_0(Q, C_n)$ so there are unique $V_m \in \mathcal{H}_o(Q(\Lambda_m))$ with $\sum_{m|n} Tr(V_m) = U$. For each V_m the eigenvalues of T acting on $V_m \otimes C$ are roots of rational primes not dividing n.

To our previous assumptions we now add the vanishing of all α invariants on U. So consider a prime $p|n$ and write $n = p^r \ell$ with $(p,\ell) = 1$ and $r > 0$. Consider

$$\sum_{k|\ell} V_k + \sum_{k|\ell} \left(\sum_{j=1}^{r} V_{p^j k} \right) = U .$$

If $\alpha(p,-)$ is applied to this equation then by IV.3.14 we find

$$\sum_{k|\ell} \alpha(p, V_{p^r k}) = \alpha(p,U) = 0$$

or

$$\sum_{k|\ell} \mathrm{cl}\, \bar{\partial}_p (V_{p^r k}) = 0 \in H^2(C_2; GR(F_p; C_n)) .$$

Appealing to the second part of IV.3.1 it again follows $\bar{\partial}_p(V_{p^r k}) = 0$. By the obvious induction procedure $\bar{\partial}_p(V_{p^j k}) = 0$ for $1 \leq j \leq r$ and $k|\ell$. For V_k with $k|\ell$ we find $\bar{\partial}_p(V_k) = 0$ just as in the case $q \mid n$.

Thus for each V_m, $\bar{\partial}(V_m) = 0 \in \mathcal{H}_0(Q(\Lambda_m)/Z(\Lambda_m))$ so $V_m \in \mathcal{H}_0(Z(\Lambda_m))$ and has vanishing multisignature. There is no torsion in $\mathcal{H}_0(Z(\Lambda_m))$ and hence $V_m = 0$. This completes the proof of (3.15). □

(3.16) <u>Corollary</u>: <u>If</u> $V \in W_0(Z, C_n)$ <u>then</u>

$$\alpha(p^s, U) = \mathrm{cl}(C_n, H^1(C_{p^s}; U))$$

<u>in</u> $H^2(C_2; GR(F_p, C_n))$. <u>Furthermore</u> U <u>is uniquely determined by</u>

1) <u>its multisignature</u>

2) <u>its</u> α-<u>invariants</u>.

Proof. In this situation $\partial(U) = 0 \in W_0(Q/Z, C_n)$ automatically. Also we can take as special $Z_{(p)}$-lattice $U \otimes Z_{(p)} = L = L^+$. Thus the α-invariants alone detect all the torsion in $W_0(Z, C_n)$. At the end of Chapter III we determined the torsion subgroup.

Obviously IV.3.16 is the focal point of the entire chapter. In the next chapter we shall make a topological application.

(3.17) Exercise: In the special case $W_0(Z, C_{p^r})$ show all α-invariants vanish. (Hint: $H^1(C_{p^s}; U)$ carries a skew-symmetric innerproduct.)

A quite similar approach applies to $W_2(Z, C_n)$. Very briefly, leaving details to the reader, we shall sketch the necessary changes. This time we follow the procedure in Chapter III more closely. Thus the isomorphism $Tr: \sum_{\substack{m|n \\ m>1}} \mathcal{H}_o(Q(\lambda_m)) \simeq W_2(Q, C_n)$ is defined as follows. If $V \in \mathcal{H}_o(Q(\lambda_m))$ has the Hermitian innerproduct $[v,w]$ the associated rational form on V is $(v,w) = tr(\Delta^{-1}[v,w])$. As in Chapter III, $\Delta = -\bar{\Delta} \in Z(\lambda_m)$ generates the different of $Q(\lambda_m)/Q$, that is, up to a unit $\Delta = u_n/n$. Again we have $\bar{\partial}(V) \in \mathcal{H}_o(Q(\lambda_m)/Z(\lambda_m))$ and $\partial \circ Tr(V) \in W_2(Q/Z, C_n)$. We propose to again compare these in $H^2(C_2; GR(Q/Z, C_n))$. In this case $W_2(Q/Z, C_n) \simeq H^2(C_2; GR(Q/Z, C_n))$. In contrast to IV.3.5 we find

(3.18) Lemma: For the skew-symmetric case, $cl(\bar{\partial}(V)) = cl(\partial \circ Tr(V))$ in $H^2(C_2; GR(Q/Z, C_n))$.

If $P \subset V$ is a $Z(\lambda_m)$-lattice for the Hermitian structure then P is at the same time a Z-lattice for the associated rational form. Furthermore $P^{\#} = P^+$. The local version will assert that for each rational prime

$$cl(\bar{\partial}_q(V)) = cl(\partial_q(V))$$

in $H^2(C_2; GR(F_q, C_n)) \simeq H^2(C_2; GR(F_q, C_\ell))$ with $\ell q^r = n$, $(\ell, q) = 1$.

If $p|n$ we can go ahead to define the α-invariants $\alpha(p^s, -)$, with $1 \leq s \leq$ multiplicity of p in n, for $W_2(Q, C_n)$. This time, for a special $Z_{(p)}$ lattice $L \subset U$ the finite form on im: $H^1(C_{p^s}; L) \longrightarrow H^1(C_{p^s}, L^+)$ happens to be symmetric, but this alters nothing. There is a difference which must be observed in the following situation. Suppose, with $n = p^r\ell$, $r > 0$ $(p, \ell) = 1$ we have selected a

$V \in \mathcal{H}_o(Z(\Lambda_{p^j k}))$ where $k|\ell$ and $1 \leq j \leq r$. The argument for IV.3.14 shows $\alpha(p^s, V) = 0$ if $j \leq r-s$, however, if $j > r-s$ we find

$$\begin{aligned} \alpha(p^s, V) &= cl(\bar{\partial}_p(V)) + a\ cl(F_p(\Lambda_k)) \\ &= cl((\partial \circ Tr)_p(V)) + a\ cl(F_p(\Lambda_k)) \end{aligned}$$

with $a = \dim_{Q(\Lambda_{p^j k})} V$.

This does not, however, interfere with our proof that an element in $W_2(Q, C_n)$ is determined by

1) its multisignature
2) its image under $\partial: W_2(Q, C_n) \longrightarrow W_2(Q/Z, C_n)$
3) its α-invariants.

The point is this. For $U \in W_2(Q, C_n)$ we uniquely choose $V_m \in \mathcal{H}_o(Q(\Lambda_m))$, $m|n$ and $m > 1$, so that $\sum Tr\ V_m = U$. As soon as we assume U has multisignature 0 we find $\dim_{Q(\Lambda_m)} V_m = 0 \pmod 2$ for all m. The rest of the argument for IV.3.15 will now apply.

Of course if $U \in W_2(Z, C_n)$ then $\alpha(p^s, U) = cl(C_n, H^1(C_{p^s}; U))$ in $H^2(C_2; GR(F_p, C_n)) \simeq H^2(C_2; GR(F_p, C_\ell))$. We also observe that, unlike the symmetric case, the α-invariants are definitely not all trivial on $W_2(Z, C_{p^r})$.

This alteration to

$$Tr: \sum_{\substack{m|n \\ m>1}} \mathcal{H}_o(Q(\Lambda_m)) \simeq W_2(Q, C_n)$$

was introduced so that we could continue to use IV.3.1 and extend IV.3 15 without changing the wording of the theorem, or the definition of the α-invariant.

4. $W_*(Z, \pi)$ and Dress' Induction Theorem

Dress' induction techniques allow us to extend results about cyclic groups of odd order to general finite groups of odd order. We discuss very briefly the relationship. To begin with we outline some of the basic properties of a Frobenius functor. Our discussion is from [Sw 3]. We refer the reader to [D3] for additional information on Frobenius, Green, and Mackey functors. A Frobenius functor $\mathcal{F}$ is a contravariant functor from the category of finite groups and inclusion maps to the category of rings and ring homomorphisms, satisfying the following additional conditions

a) If $i: \pi' \longrightarrow \pi$ is an inclusion of finite groups then there is a group homomorphism

$$i_*: \mathcal{F}(\pi') \longrightarrow \mathcal{F}(\pi)$$

b) If $i^*: \mathcal{F}(\pi) \longrightarrow \mathcal{F}(\pi')$ then i^* and i_* are related by

$$x \cdot i_* y = i_*(i^* x \cdot y)$$

where $\cdot$ indicates multiplication in the appropriate ring, $x \in \mathcal{F}(\pi)$, $y \in \mathcal{F}(\pi')$.

Examples

1) $\mathcal{F}(\pi) = R_C(\pi)$ is a Frobenius functor. i^* is the reduction homomorphism and i_* is the induction homomorphism.
2) $\mathcal{F}(\pi) = W_*(Q,\pi)$ is a Frobenius functor, i^* is reduction, i_* is induction.
3) $\mathcal{F}(\pi) = W_*(Z,\pi)$ is a Frobenius functor.
4) $\mathcal{F}(\pi) = \mathcal{O}_*(\pi)$, the cobordism ring of orientation preserving actions of π on closed manifolds is a Frobenius functor. Condition (b) is verified in [C-F].

Let $\mathcal{C}$ be a collection of subgroups of π. Let $D_{\mathcal{F}}(\mathcal{C})$ be the subgroup of $\mathcal{F}(\pi)$ generated by $i_*(\pi')$ for $\pi' \in \mathcal{C}$ and $\tilde{D}(\mathcal{C}) = \ker\{\oplus i^*: \mathcal{F}(\pi) \to \bigoplus_{\pi' \in \mathcal{C}} \mathcal{F}(\pi')\}$.

(4.1) Lemma: a) $D_{\mathcal{F}}(\mathcal{C})$ is an ideal of $\mathcal{F}(\pi)$.

b) $\tilde{D}_{\mathcal{F}}(\mathcal{C})$ is naturally a module over $\mathcal{F}(\pi)/D_{\mathcal{F}}(\mathcal{C})$.

Proof. a) If $y \in D_{\mathcal{F}}(\mathcal{C})$ then $y = \sum_{\pi' \in \mathcal{C}} i_* y'$ so that $x \cdot y = \sum x \cdot i_* y' = \sum i_*(i^* x \cdot y') \in D_{\mathcal{F}}(\mathcal{C})$.

b) We need to show that $D_{\mathcal{F}}(\mathcal{C})$ annihilates $\tilde{D}_{\mathcal{F}}(\mathcal{C})$. For y as above and $x \in \tilde{D}_{\mathcal{F}}(\mathcal{C})$ we get $x \cdot y = \sum i_*(i^* x \cdot y') = 0$ because $i^* x = 0$ for all i^*. □

(4.2) Lemma: *Consider* $\mathcal{F}(\pi) = W_*(Q,\pi)$ *or* $W_*(Z,\pi)$ *and let* $\mathcal{C}$ *be the collection of cyclic subgroups of* π. *Then* $\tilde{D}_W(\mathcal{C})$ *is a torsion group and* $4x = 0$ *for all* $x \in \tilde{D}_W(\mathcal{C})$.

Proof. $W_*(Q,\pi)$ modulo torsion injects into $R_C(\pi)$ and it is well known that $\tilde{D}_{R_C}(\mathcal{C}) = \{0\}$. Therefore $\tilde{D}_W(\mathcal{C})$ is a torsion group but the torsion subgroup of $W_*(Q,\pi)$ has exponent four. □

We present without proof the theorem of Dress.

(4.3) Theorem (Dress): *If*   *denotes the unit in the ring* $W_*(Q,\pi)$ (*or* $W_*(Z,\pi)$) *then*

a) $|\pi| \cdot 1 \in D_W(\mathcal{C})$ *when* $|\pi|$ *is odd*

b) *If* $|\pi| = 2^k \cdot |\pi|'$ *where* $|\pi|'$ *is odd then* $4|\pi|' \cdot 1 \in D_W(\mathcal{C})$ *in general*. □

(4.4) Corollary (Dress): *If* $|\pi|$ *is odd then* $\tilde{D}_W(\mathcal{C}) = \{0\}$. *In particular this means that for an odd order group the equivariant Witt class is determined by its restriction to cyclic subgroups.*

Proof. $\tilde{D}_W(\mathcal{C})$ is a 2-torsion group and by IV.4.1 it is a module over $W_*(Q,\pi)/D_W(\mathcal{C})$. This last ring has odd order by IV.4.3(a) so that $\tilde{D}_W(\mathcal{C}) = \{0\}$. □

The results of IV.4.3 are surprising. It is not easy to see how a multiple of 1 occurs in the image of the induction homomorphism when π is *not* cyclic. We present two simple examples for the reader: The case when $\pi = C_3 \times C_3$ and when π is the non-abelian group of order 21.

$\pi = C_3 \times C_3$: π has four non-trivial cyclic subgroups of order 3. Denote them by $\{H_i | 1 \leq i \leq 4\}$ and their quotients by H^i. In $W_0(Z, H^i)$, there is the regular representation $Z(H^i)$ with the standard innerproduct. Let V^i be the pullback to $C_3 \times C_3$ of this form. V^i is induced by $1 \in W_0(Z, H_i)$. Let U be the regular representation which is clearly in the image of the induction homomorphism. Then

$$3 \cdot 1 = [V^1] + [V^2] + [V^3] - [U] .$$

$|\pi| = 21$, π non-abelian: There are six subgroups of order 3 and all of them are conjugate. Let H denote one of these subgroups. Let K be the subgroup of order 7. Let V be induced by $1 \in W_0(Z, H)$, U induced by $1 \in W(Z, K)$ and R the inflation of $1 \in W(Z)$. Then

$$3 \cdot 1 = 3V + U - R .$$

(Checking this makes one appreciate the complexities of IV.4.3.)

(4.5) <u>Corollary</u>: $V \in W_0(Z, \pi)$ is uniquely determined by its total signature and $\{cl(H^1(C_{p^s}; V)) \in H^2(C_2; GR(F_p, C_\ell)) | T \in \pi$ is an element of order $n = p^r \ell$, $(p, \ell) = 1$, $1 \leq s \leq r$, and C_n acts on V via T.$\}$ □

Let us give one more example of the application of Dress' induction in the context of these notes. We recall that there is a ring homomorphism

$$W_0(Z, \pi) \longrightarrow H^2(C_2; GR(Z, \pi))$$

which preserves identity elements. We use this to make $H^2(C_2; GR(Z, \pi))$ into a $W_0(Z, \pi)$-module.

If $\pi' \subset \pi$ is a subgroup then there are defined restriction homomorphisms so that the diagram

$$\begin{array}{ccc} W_0(Z,\pi) & \longrightarrow & H^2(C_2, GR(Z,\pi)) \\ \downarrow r_\pi^{\pi'} & & \downarrow R_\pi^{\pi'} \\ W_0(Z,\pi') & \longrightarrow & H^2(C_2; GR(Z,\pi')) \end{array}$$

is commutative. There is also the induction homomorphism $i_{\pi'}^{\pi}: W_0(Z,\pi') \longrightarrow W_0(Z,\pi)$. Because we shall use it in the next chapter let us show there is a corresponding induction homomorphism

$$I_{\pi'}^{\pi}: H^2(C_2; GR(Z,\pi')) \longrightarrow H^2(C_2; GR(Z,\pi)) .$$

If (π', V) is an integral representation of π' on a finite abelian group then $Z(\pi) \otimes_{Z(\pi')} V$ is the induced representation of π. We need to show there is a left $Z(\pi)$-module isomorphism

$$Z(\pi) \otimes_{Z(\pi')} \mathrm{Hom}_Z(V,Z) \simeq \mathrm{Hom}_Z(Z(\pi) \otimes_{Z(\pi')} V, Z) .$$

This will say, at the level of $GR(Z,\pi') \longrightarrow GR(Z,\pi)$, the induction will commute with the duality involution and hence $I_{\pi'}^{\pi}: H^2(C_2; GR(Z,\pi')) \longrightarrow H^2(C_2; GR(Z,\pi))$ is induced as desired.

Now on $\mathrm{Hom}_Z(V,Z)$ there are right and left $Z(\pi')$-module structures given by

$$(\eta h)(v) \equiv \eta(hv)$$
$$(h\eta)(v) \equiv \eta(h^{-1}v) = (\eta h^{-1})(v)$$

where $h \in \pi'$, $\eta \in \mathrm{Hom}_Z(V,Z)$ and $v \in V$. We regard $Z(\pi)$ as a right $Z(\pi')$-module.

According to [C-E, II.5.2'] there is a natural isomorphism $\mathrm{Hom}_{Z(\pi')}(Z(\pi), \mathrm{Hom}_Z(V,Z)) \simeq \mathrm{Hom}_Z(Z(\pi) \otimes_{Z(\pi')} V, Z)$. By naturality this is an isomorphism of left $Z(\pi)$-modules. Using the left $Z(\pi')$-module structure on $\mathrm{Hom}_Z(V,Z)$ we can also form $Z(\pi) \otimes_{Z(\pi')} \mathrm{Hom}_Z(V,Z)$ and we claim there is a $Z(\pi)$-module isomorphism $\mathrm{Hom}_{Z(\pi')}(Z(\pi), \mathrm{Hom}_Z(V,Z)) \simeq Z(\pi) \otimes_{Z(\pi')} \mathrm{Hom}_Z(V,Z)$.

A basis for $Z(\pi)$ as a free right $Z(\pi')$ is given by choosing coset representatives $g_0 = e, g_1, \ldots, g_k$ of π' in π. Then every element of $\mathrm{Hom}_{Z(\pi')}(Z(\pi), \mathrm{Hom}_Z(V,Z))$ is uniquely determined by its values on the g_j and every element in $Z(\pi) \otimes_{Z(\pi')} \mathrm{Hom}_Z(V,Z)$ can be unique written as $\sum_{j=0}^{k} g_j \otimes \eta_j$ with $\eta_j \in \mathrm{Hom}_Z(V,Z)$. The isomorphism in question will send $\psi \in \mathrm{Hom}_{Z(\pi')}(Z(\pi), \mathrm{Hom}_Z(V,Z))$ into $\sum_{j=0}^{k} g_j \otimes \Psi(g_j)$.

To see this preserves $Z(\pi)$-module structures we note that for each $g \in \pi$ there is a unique permutation, denoted by $g_j \longrightarrow g_{f(j)}$, of the set of coset representatives and a unique choice of elements $h_j \in \pi'$ so that

$$g^{-1} g_j = g_{f(j)} h_j^{-1} .$$

Then the action of g on ψ is

$$(g\psi)(g_j) = \psi(g^{-1} g_j) = \psi(g_{f(j)} h_j^{-1}) = \psi(g_{f(j)}) h_j^{-1} .$$

Thus

$$\begin{aligned}(g\psi) \quad & \sum_{j=0}^{k} g_j \otimes \psi(g_{f(j)}) h_j^{-1} \\ &= \sum_{j=0}^{k} g_j h_j \otimes \psi(g_{f(j)}) \\ &= \sum_{j=0}^{k} g g_{f(j)} \otimes \psi(g_{f(j)}) \\ &= g\Big(\sum_{j=0}^{k} g_j \otimes \psi(g_j)\Big) .\end{aligned}$$

Thus

$$Z(\pi) \otimes_{Z(\pi')} \mathrm{Hom}_Z(V,Z) \simeq \mathrm{Hom}_Z(Z(\pi) \otimes_{Z(\pi')} V, Z)$$

as $Z(\pi)$-modules. This means that at the level of Grothendieck groups induction commutes with dualizing. The resulting diagram

$$\begin{array}{ccc} W_0(Z,\pi') & \longrightarrow & H^2(C_2;GR(Z,\pi')) \\ \downarrow i^{\pi}_{\pi'} & & \downarrow I^{\pi}_{\pi'} \\ W_0(Z,\pi) & \longrightarrow & H^2(C_2;GR(Z,\pi)) \end{array}$$

is also commutative. We shall leave the verification of the following to the reader.

(4.6) Theorem: The functor $H^2(\pi' GR(Z,\pi))$ is itself a Green functor and furthermore it is a module over the Green functor $W_0(Z,\pi)$.

If π has odd order then some odd multiple of $1 \in W_0(Z,\pi)$ lies in the $\Sigma W_0(Z,\gamma) \xrightarrow{i^{\pi}_{\gamma}} W_0(Z,\pi)$. Since $H^2(C_2;GR(Z,\pi))$ is an elementary abelian 2-group we can draw

(4.7) Corollary: If π is a finite odd order group then

$$\Sigma H^2(C_2;GR(Z,\gamma)) \xrightarrow{\oplus I^{\pi}_{\gamma}} H^2(C_2;GR(Z,\pi)$$

is an epimorphism and

$$H^2(C_2;GR(Z,\pi)) \xrightarrow{\oplus R^{\gamma}_{\pi}} \Sigma H^2(C_2;GR(Z,\gamma))$$

is a monomorphism. The sum is taken over all cyclic subgroups $\gamma \in \pi$.

There are other functors to which the above considerations apply. For example if p is a rational prime then reduction mod p is used to make $W_0(F_p,\pi)$ into a $W_0(Z,\pi)$-module. Of course as Dress shows $W_0(F_p,\pi)$ is itself a Green functor. Since $W_0(F_p,\pi)$ is a finite abelian 2-group the corollary IV.4.4 will apply if π has odd order. Still another example is $H^2(C_2;GR(F_p,\pi)$. In this chapter we analysed

$$W_0(F_p,C_n) \longrightarrow H^2(C_2;GR(F_p,C_n)).$$

What is the analogue if π has odd order?

There is yet another situation in which duality defines an involution. Let $\tilde{K}_0(Z[\pi])$ be the projective class group associated with the integral group ring $Z[\pi]$. If P is a finitely generated $Z[\pi]$-projective then its dual is $\mathrm{Hom}_{Z[\pi]}(P, Z[\pi])$ with the usual left $Z[\pi]$ module structure. This will give us

$$H^*(C_2;\tilde{K}_0(Z[\pi])) = H^1(C_2;\tilde{K}_0(Z[\pi])) \oplus H^2(C_2;\tilde{K}_0(Z[\pi])) .$$

In this case $H^*(C_2;\tilde{K}_0(Z[\pi]))$ is not a Green functor but rather following the suggestions found both in [Sw1,D3] we regard it as a module over $W_0(Z,\pi)$. In other words, if $V \in W_0(Z,\pi)$ and P is a projective $Z[\pi]$-module then $V \otimes_Z P$ with the diagonal action of π is $Z[\pi]$-projective. Now $H^*(C_2;\tilde{K}_0(Z[\pi]))$ is a Mackey functor and a module over $W_0(Z,\pi)$.

If π has odd order then some odd multiple of $1 \in W_0(Z,\pi)$ lies in the image of $\sum W_0(Z,\gamma) \longrightarrow W_0(Z,\pi)$. This is used to show the corollary IV.4.4 applies to $H^*(C_2;\tilde{K}_0(Z[\pi]))$.

In chapter 6 we shall discuss an exposition of a special case, π cyclic of odd prime power order, of a theorem announced by Bak [B]. However it will follow immediately that if π is a p-group then $0 \longrightarrow H^1(C_2;\tilde{K}_0(Z[\pi])) \longrightarrow L_0^h(\pi) \longrightarrow L_0^p(\pi)$ $H^2(C_2;\tilde{K}_0(Z[\pi])) \longrightarrow 0$ is exact. Bak has shown this for odd order groups and we presume his approach is to verify it first in the cyclic case and then employ Dress' induction as we have.

To close the discussion of Dress' induction we propose a question. Consider $W_0(Q/Z,\pi)$. This is surely a $W_0(Z,\pi)$ module as is $W_0(Q,\pi)$ for that matter. Furthermore it is easy to see that the boundary homomorphism

$$\partial\colon W_0(Q,\pi) \longrightarrow W_0(Q/Z,\pi)$$

is a $W_0(Z,\pi)$-module homomorphism. We ask

1. If C_n is a cyclic group of odd order is

$\partial : W_0(Q,C_n) \longrightarrow W_0(Q/Z,C_n)$ an epimorphism?

2. If the answer to 1 is yes then will this also follow for π an odd order group?

   Remember $W_0(Q/Z,\pi) \simeq \sum_p W_0(F_p,\pi)$.

Chapter V

Torsion Signature Theorem

1. Induction and restriction in $H^2(C_2;GR(F_p,C_\ell))$

Suppose that ℓ and p are odd, p is a prime and $(p,\ell) = 1$. For subgroups $C_{k'} \subset C_k \subset C_\ell$ there is the homomorphism

$$R_k^{k'} : H^2(C_2;GR(F_p,C_k)) \longrightarrow H^2(C_2;GR(F_p,C_{k'}))$$

obtained by restricting the action to that of the subgroup $C_{k'}$. There is also the induction homomorphism

$$I_{k'}^k : H^2(C_2;GR(F_p,C_{k'})) \longrightarrow H^2(C_2;GR(F_p,C_k))$$

which to $cl(V)$ assigns $cl(F_p[C_k] \otimes_{F_p[C_{k'}]} V)$. We note that $R_k^{k'} \circ I_{k'}^k\, cl(V) = (k/k')cl(V) = cl(V)$ since k/k' is odd. Therefore

$$R_k^{k'} \circ I_{k'}^k = \text{identity} .$$

If $C_{k''} \subset C_{k'} \subset C_k$ then there are the usual associativity formulas

$$R_k^{k''} = R_{k'}^{k''} \circ R_k^{k'}$$

$$I_{k''}^k = I_{k'}^k \circ I_{k''}^{k'} .$$

There is still another relation following from Curtis-Reiner [Cu-R].

(1.1) Lemma: If C_k, $C_{k'}$ are any two subgroups of C_ℓ then with $C_{k''} = C_k \cap C_{k'}$

$$R^{k'}_{\ell} \circ I^{\ell}_{k} = I^{k'}_{k''} \circ R^{k''}_{k} . \quad \square$$

We are concerned with the subgroups $E_k \subset H^2(C_2;GR(F_p,C_k))$ which is defined to be the image of

$$\sum_{\substack{k'|k \\ k'\neq k}} H^2(C_2;GR(F_p,C_{k'})) \xrightarrow{\oplus I^k_{k'}} H^2(C_2;GR(F_p,C_k)) .$$

In particular there is $E_\ell \subset H^2(C_2;GR(F_p,C_\ell))$.

(1.2) Definition: If $cl(V) \in E_\ell$ then a distribution for $cl(V)$ is a function d which to each proper divisor k of ℓ assigns an element $d(k) \in H^2(C_2;GR(F_p,C_k))$ so that $\sum_{\substack{k|\ell \\ k\neq \ell}} I^{\ell}_{k} d(k) = cl(V)$. The distribution is tight if and only if $d(k) \in E_k$ implies $d(k) = 0$.

Consider a distribution and suppose that for some proper divisor k_0/ℓ, $0 \neq d(k_0) \in E_{k_0}$. Then there is a subdistribution δ which to every proper divisor k'/k_0 assigns $\delta(k')$ in $H^2(C_2;GR(F_p,C_{k'}))$ so that

$$\sum_{\substack{k'|k_0 \\ k'\neq k_0}} I^{k_0}_{k'}(\delta(k')) = d(k_0) .$$

We then alter the original distribution into

$$D(k) = d(k), \; k \nmid k_0$$

$$D(k_0) = 0$$

$$D(k) = d(k) + \delta(k), \; k \mid k_0, \; k \neq k_0 .$$

In view of the associative law for induction this is again a distribution for $cl(V)$.

(1.3) Lemma: For any element in E_ℓ there is a tight distribution.

Proof. In this argument the proper divisors of ℓ are numbered off $k_1, k_2 \ldots$, so that if $k_i | k_j$ then $i > j$. Fix an index i. Assume by induction there is a distribution such that if $j < i$ then $d(k_j) = 0$ or $d(k_j) \notin E_{k_j}$. If $0 \neq d(k_i) \in E_{k_i}$ we make the alteration of the distribution as described before. If $j \neq i$ this does not change $d(k_j)$ unless k_j is a proper divisor of k_i, but in that case $j > i$. Thus the induction proceeds. □

For every divisor $k|\ell$ let $\bar{R}^k_\ell$ denote the composite homomorphism

$$H^2(C_2; GR(F_p, C_\ell)) \xrightarrow{R^k_\ell} H^2(C_2; GR(F_p, C_k)) \longrightarrow H^2(C_2; GR(F_p, C_k))/E_k .$$

We now claim

(1.4) Lemma: The sum

$$H^2(C_2; GR(F_p, C_\ell)) \xrightarrow{\bar{R}^k_\ell} \bigoplus_{k|\ell} H^2(C_2; GR(F_p, C_k))/E_k$$

is an injection.

Proof. For this argument we again number off the proper divisors of ℓ, $k_1, k_2 \ldots$, so that if $k_i | k_j$ then $j < i$. Suppose $cl(V)$ belongs to the kernel. Then by definition $cl(V) \in E_\ell$. Suppose $k_j \longrightarrow cl(V_{k_j}) \in H^2(C_2; GR(F_p, C_{k_j}))$ is a tight distribution for $cl(V)$. By way of induction we suppose, for a fixed index i, we have shown $cl(V_{k_j}) = 0$ for all $j < i$. This includes those k_j of which k_i is a divisor. Now we consider

$$R^{k_i}_\ell cl(V) = cl(V_{k_i}) + \sum_{j>i} R^{k_i}_\ell \circ I^\ell_{k_j} cl(V_{k_j}) .$$

We apply (V.1.1) to each term in the sum using $k = k_j$ and $k' = k_i$ so

$$R_{\ell}^{k_i} \circ I_{k_j}^{\ell} \, cl(V_{k_j}) = I_{k''}^{k_i} \circ R_{k_j}^{k''} \, cl(V_{k_j}) \; .$$

Since k_i does not divide k_j

$$C_{k''} = C_{k_j} \cap C_{k_i}$$

is not C_{k_i}. Hence every term in the sum belongs to E_{k_i}. By assumption $\bar{R}_{\ell}^{k_i} cl(V) = 0$ so that $R_{\ell}^{k_i} cl(V) \in E_{k_i}$. Therefore $cl(V_{k_i})$ lies in E_{k_i} and by the tightness assumption $cl(V_{k_i}) = 0$. □

To continue, it is the quotient group $H^2(C_2;GR(F_p,C_{\ell}))/E_{\ell}$ with which we are mainly concerned. Suppose $k|\ell$ then there is the natural quotient homomorphism

$$\nu_k : C_{\ell} \longrightarrow \tilde{C}_k \; .$$

We use $\tilde{C}_k$ to remind the reader that we mean the quotient group. Surely there is induced

$$\nu_k^* : H^2(C_2;GR(F_p,\tilde{C}_k)) \longrightarrow H^2(C_2;GR(F_p,C_{\ell})) \; .$$

Now suppose $\tilde{C}_{k'} \subset \tilde{C}_k$, then there is the subgroup $\nu_k^{-1}(\tilde{C}_{k'}) = C_{k'k''} \subset C_{\ell}$ where $k'' = k/k'$ is the order of the kernel of ν_k. We note two commutative diagrams. First

$$\begin{array}{ccc} H^2(C_2;GR(F_p,\tilde{C}_k)) & \xrightarrow{\nu_k^*} & H^2(C_2;GR(F_p,C_{\ell})) \\ \Big\downarrow R_k^{k'} & & \Big\downarrow R_{\ell}^{k'k''} \\ H^2(C_2;GR(F_p,\tilde{C}_{k'})) & \xrightarrow{\nu_{k'}^*} & H^2(C_2;GR(F_p,C_{k'k''})) \end{array}$$

and also

$$\begin{array}{ccc} H^2(C_2;GR(F_p,\widetilde{C}_{k'})) & \xrightarrow{\nu^*_{k'}} & H^2(C_2;GR(F_p,C_{k'k''})) \\ \downarrow \widetilde{I}^k_{k'} & & \downarrow I^{\ell}_{k'k''} \\ H^2(C_2;GR(F_p,\widetilde{C}_k)) & \xrightarrow{\nu^*_k} & H^2(C_2;GR(F_p,C_\ell)) \end{array}$$

In particular the second diagram allows us to conclude there is an induced

$$\overline{\nu}^*_k:\ H^2(C_2;GR(F_p,\widetilde{C}_k))/E_k \longrightarrow H^2(C_2;GR(F_p,C_\ell))/E_\ell \ .$$

Let us draw a simple illustrative corollary.

(1.5) <u>Lemma</u>: <u>If</u> $C_{k''} = \ker(\nu_k)$ <u>then</u> $\nu^*_k\ cl(F_p[\widetilde{C}_{p^k}]) = I^{\ell}_{k''}(cl(F_p))$. □

We observe $cl(F_p[\widetilde{C}_k])$ is the image under $\widetilde{I}^k_1$: $H^2(C_2;GR(F_p,\widetilde{C}_1))$. $H^2(C_2\ GR(F_p,\widetilde{C}_k))$ of $cl(F_p)$ and apply the commutative diagram for induction

(1.6) <u>Corollary</u>: <u>The image of the Burnside ring under the composite</u>

$$Burn(C_\ell) \longrightarrow H^2(C_2;GR(F_p,C_\ell)) \longrightarrow H^2(C_2;GR(F_p,C_\ell))/E_\ell$$

<u>is generated by</u> $\overline{cl}(F_p)$, <u>the image of the trivial one dimensional representation of</u> C_ℓ. □

We want to know exactly when $\overline{cl}(F_p) \neq 0$. If ℓ has prime factorization $q^{r_1} \ldots q_s^{r_s}$ we put $k_0 = q_1 \ldots q_s$. There is the ring of integers $Z(\lambda_{k_0})$ in the k_0-cyclotomic extension. Then $F_p(\lambda_{k_0}) = Z(\lambda_{k_0})/(p)$ and we have $cl(F_p(\lambda_{k_0})) \in H^2(C_2;GR(F_p,C_\ell))$.

(1.7) <u>Lemma</u>: <u>Modulo the subgroup</u> E_ℓ

$$\overline{cl}(F_p) = \overline{cl}(F_p(\lambda_{k_0})) \ .$$

Proof. We need only consider this relation in $H^2(C_2;GR(F_p;C_{k_0}))/E_{k_0}$ because there is $\nu_{k_0} : C_\ell \longrightarrow \tilde{C}_{k_0}$ and if we can establish the relation in $\tilde{C}_{k_0}$ it will follow for C_ℓ by applying $\vec{\nu}^*_{k_0}$.

If $k' | k_0$ is a proper divisor then $\overline{cl}(F_p[\tilde{C}_{k'}]) = 0 \in H^2(C_2; GR(F_p, C_{k_0}))/E_{k_0}$ as previously noted. Furthermore

$$cl(F_p[\tilde{C}_{k'}]) = \sum_{j|k'} cl(F_p(\Lambda_j)) .$$

Thus we can write

$$\overline{cl}(F_p) = \sum_{k'|k_0} \overline{cl}(F_p(\tilde{C}_{k'})) = \sum_{k'|k_0} \Big(\sum_{j|k'} \overline{cl}(F_p(\Lambda_j)\Big) .$$

Note $F_p(\Lambda_j)$ is a C_{k_0}-module via $C_{k_0} \longrightarrow \tilde{C}_{k'} \longrightarrow \tilde{C}_j$. In this last sum if $j < k_0$ the term $F_p(\Lambda_j)$ appears an even number of times while $F_p(\Lambda_{k_0})$ appears but once. Hence $\overline{cl}(F_p) = \overline{cl}\, F_p(\Lambda_{k_0})$. □

Now we return to

$$\nu_{k_0} : C_\ell \longrightarrow \tilde{C}_{k_0} .$$

We want to show that if k' is a proper divisor then $R_\ell^{k'}\, cl(F_p(\Lambda_{k_0})) = 0 \in H^2(C_2; GR(F_p, C_{k'}))$. Here we observe that $\tilde{C}_{k_1} = \nu_k(C_{k'}) \subset \tilde{C}_{k_0}$ <u>is also a proper subgroup</u>. Now $\tilde{R}_{k_0}^{k_1}\, cl(F_p(\Lambda_{k_0})) = \varphi(k_0/k_1) cl(F_p(\Lambda_{k_1}))$ where φ is the Euler φ-function. A simple count of dimensions over F_p shows this. But $\varphi(k_0/k_1)$ is even so $\tilde{R}_{k_0}^{k_1}\, cl(F_p(\Lambda_{k_0})) = 0$.

Finally we consider

$$\begin{array}{ccc}
H^2(C_2;GR(F_p,\tilde{C}_{k_0})) & \xrightarrow{\nu_k^*} & H^2(C_2;GR(F_p,C_\ell)) \\
\Big\downarrow \tilde{R}^{k_1}_{k_0} & & \Big\downarrow R^{k_1k''}_{k} \\
H^2(C_2;GR(F_p,C_{k_1})) & \xrightarrow{\nu^*} & H^2(C_2;GR(F_p,\nu_k^{-1}(\tilde{C}_{k_1}))) \\
 & & \Big\downarrow R^{k'}_{k_1k''} \\
 & & H^2(C_2;GR(F_p,C_{k'}))
\end{array}$$

to see $R^{k'}_{\ell}\ cl(F_p(\Lambda_{k_0})) = 0$ if k' is a proper divisor of ℓ. Thus according to (V.1.4) $cl(F_p(\Lambda_{k_0})) \neq 0$ if and only if $\overline{cl}(F_p(\Lambda_{k_0})) = \overline{cl}(F_p) \neq 0$.

The criterion for non-vanishing lies in the following consideration. The prime p generates a multiplicative subgroup of Z_ℓ^*, the group of units in the ring Z_ℓ. Denote this subgroup by $gr(p) \subset Z_\ell^*$.

(1.8) Lemma: The element $\overline{cl}(F_p) \neq 0 \in H^2(C_2;GR(F_p,C_\ell))/E_\ell$ if and only if $-1 \in gr(p)$.

Proof. Actually we need to show $cl(F_p(\Lambda_{k_0})) \neq 0 \in H^2(C_2,GR(F_p,C_\ell))$. The criterion is from chapter IV. If $\Phi_{k_0}(t)$ is the k_0-cyclotomic polynomial then $\Phi^*_{k_0}(t) = \Phi_{k_0}(t)$. When reduced mod p this factors into distinct irreducibles

$$\Phi_{k_0}(t) = f_1(t) \ldots f_r(t)$$

all distinct. But $cl(F_p(\Lambda_{k_0})) \neq 0$ if and only if $f_j^*(t) = f_j(t)$, $1 \leq j \leq r$ by IV.2.7 which happens if and only if $-1 \in gr(p)$. □

2. The Inductive Step

For an orientation preserving diffeomorphism of period n on a compact, connected, oriented manifold M^{2m}, the torsion signatures are defined for each prime $p|n$ so that $n = p^r \cdot \ell$, $(p,\ell) = 1$, and $-1 \in gr(p) \subset Z_\ell^*$. For such a p, let

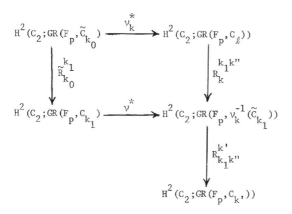

$V = H^m(M;Z)/TOR$. By Lemma II.1.2, $H^i(C_{p^r};V)$ admits a C_ℓ-invariant nonsingular finite form. This form is symmetric if $i \equiv m \bmod 2$ and skew symmetric if $i \not\equiv m \bmod 2$. We therefore have elements

$$[C_\ell, H^m(C_{p^r};V)] \in W_0(F_p, C_\ell)$$

$$[C_\ell, H^{m+1}(C_{p^r};V)] \in W_2(F_p, C_\ell) .$$

The image of $H^i(C_{p^r}, V)$ in $H^2(C_2;GR(\mathbb{F}_p, (\ell))$ is denoted $\overline{cl}(H^i(C_{p^r};V))$ for $i = 1$ or 2.

(2.1) Definition: The torsion signatures are

$$f(T,p) = \overline{cl}(H^1(C_{p^r};V)) - \overline{cl}(H^1(C_{p^{r-1}};V))$$

$$g(T,p) = \overline{cl}(H^2(C_{p^r};V)) - \overline{cl}(H^2(C_{p^{r-1}};V))$$

as elements of $H^2(C_2;GR(Z_p, C_\ell))/E_\ell$.

The computation of $f(T,p)$ and $g(T,p)$ in terms of local data will be by induction. The proof starts by verifying an inductive step. Suppose $r \geq 2$ and the subgroup $C_p \subset C_{p^r} \times C_\ell$ acts trivially on M. Then there is an induced action of the quotient group $\tilde{C}_{n/p} = C_n/C_p = \tilde{C}_{p^{r-1}} \times C_\ell$. Let $\tilde{T}$ be the image of the generator T in $\tilde{C}_{n/p}$. Let f and $\tilde{f}$ denote $f(T,p)$ and $f(\tilde{T},p)$ respectively and similarly for g and $\tilde{g}$.

(2.2) Theorem: $f = \tilde{f}$ and $g = \tilde{g}$.

Proof. If A is a free abelian C_n-module such that $C_p \subset C_n$ acts trivially, one easily checks that $H^1(C_{p^r};A)$ is C_ℓ-equivariant by isomorphic to $H^1(\tilde{C}_{p^{r-1}};A)$. It immediately follows that $f = \tilde{f}$.

To show $g = \tilde{g}$ consider

$$f+g = H^*(C_{p^r};H^m(M;Z)/TOR) - H^*(C_{p^{r-1}};H^m(M;Z)/TOR) \ .$$

By Proposition IV.1.2

$$cl(H^*(C_{p^r};H^m(M;Z)/TOR)) = cl(H^*(C_{p^r};C^*(M;Z))) \ .$$

$C^*(M;Z)$ is a direct sum of $P_1 \otimes_Z P_2$ where P_1 and P_2 are permutation representations of C_{p^r} and C_ℓ respectively and $C_p \subset C_{p^r}$ acts trivially. We can identify P_1 and P_2 with group rings $Z[\tilde{C}_{p^s}]$ and $Z[\tilde{C}_k]$ where $\tilde{C}_{p^s}$ and $\tilde{C}_k$ are quotient groups of C_{p^r} and C_ℓ and because C_p acts trivially, $s < r$.

Now $H^*(C_{p^r};P_1 \otimes P_2) = Z_{p^{r-s}} \otimes P_2$ so that

$$\overline{cl}(H^*(C_{p^r};P_1 \otimes P_2)) = \begin{cases} 0 & k > 1 \\ \overline{cl}(Z_{p^{r-s}}) & k = 1 \ . \end{cases}$$

Thus

$$\overline{cl}(H^*(C_{p^r};P_1 \otimes P_2)) - \overline{cl}(H^*(C_{p^{r-1}};P_1 \otimes P_2)) = \begin{cases} 0 & \text{if} \quad k > 1 \\ \overline{cl}(F_p) & \text{if} \quad k = 1 \ . \end{cases}$$

This means that

$$\begin{aligned} f+g &= \overline{cl}(H^*(C_{p^r};H^m(M;Z)/TOR) - \overline{cl}\ H^*(C_{p^{r-1}};H^m(M;Z)/TOR) \\ &= \overline{cl}(H^*(C_{p^r};C^*(M))) - \overline{cl}(H^*(C_{p^{r-1}};C^*(M))) \end{aligned}$$

by Proposition IV.1.2. By the computation above

$$\begin{aligned} &\overline{cl}(H^*(C_{p^r};C^*(M))) - \overline{cl}(H^*(C_{p^{r-1}};C^*(M))) \\ &= \overline{cl}(F_p \otimes C^*(Fix(T))) \ . \end{aligned}$$

Similarly

$$\tilde{f} + \tilde{g} = \overline{cl}(F_p \otimes C^*(Fix(\tilde{T}))) = \overline{cl}(F_p \otimes C^*(Fix(T))) = f + g . \quad \square$$

3. <u>Differentiable Actions of</u> C_n, n odd

For any closed oriented Y^{2k-1} denote by G the torsion subgroup of $H^k(Y;Z)$. If $x, y \in G$, then $x = \beta(z)$ for some $z \in H^{k-1}(Y;Q/Z)$ where β is the Bockstein homomorphism for the coefficient sequence $0 \to Z \to Q \to Q/Z \to 0$. Using the pairing $Z \times Q/Z \to Q/Z$ define $\mu(x,y) = \langle z \cup y, [Y] \rangle \in Q/Z$. μ is the linking form of Y, denoted $Lk(Y)$. μ represents an element of $W_0(Q/Z)$ if k is even and $W_2(Q/Z)$ if k is odd. If Y admits an orientation preserving action of π, then we consider $Lk(Y) \in W_0(Q/Z,\pi)$ for k even and $Lk(Y) \in W_2(Q/Z,\pi)$ for k odd.

If X^{2k} is a compact oriented manifold and π acts on X as a group of orientation preserving diffeomorphisms, then there is a π invariant rational form defined on

$$V = Im(H^k(X,\partial X;Q) \xrightarrow{j^*} H^k(X;Q))$$

by $b(x,y) = \langle j^{*-1}(x) \cup y, [X] \rangle$.

Note: $\partial(\pi,V,b) \in W_*(Q/Z,\pi)$.

(3.1) <u>Lemma</u>: <u>If</u> $(Y^{2k-1},\pi) = \partial(X^{2k},\pi)$, <u>then</u> $Lk(Y) = -\partial(\pi,V,b)$.

Proof. The proof is preceded by another lemma. Let $i: Y \to X$ be the inclusion and define the following subgroups of G:

$$H = \{x \in G | x = i^*\xi \quad \text{for some} \quad \xi \in H^k(X;Z)\}$$

and $$K = \{x \in G | x = i^*\xi \quad \text{for some} \quad \xi \in TOR(H^k(X;Z)))\} .$$

Applying Poincare duality one finds that $G/H \approx K$.

(3.2) Lemma: $K^{\perp} = H$ and $Lk(Y) = \langle \mu, G \rangle = \langle \mu', H/K \rangle$.

Proof. Since $G/K^{\perp} \approx \mathrm{Hom}(K, Q/Z)$, if $H \subset K^{\perp}$, the epimorphism $G/H \qquad G/K^{\perp}$ must be an isomorphism because $|G/H| = |K| = |\mathrm{Hom}(K,Q/Z)| = |G/K^{\perp}|$ and therefore $H = K^{\perp}$. Take $x \in K$ and $y \in H$.

In the diagram

$$\begin{array}{ccccc} H^{k-1}(X;Q/Z) & \xrightarrow{\ i^{*}\ } & H^{k-1}(Y;Q/Z) & \xrightarrow{\ \delta\ } & H^{k}(X,Y;Q/Z) \\ \big\downarrow{\scriptstyle\beta} & & \big\downarrow & & \\ H^{k}(X;Z) & \longrightarrow & H^{k}(Y;Z) & & \end{array}$$

there exist $\xi, \eta \in H^{k}(X;Z)$ with $i^{*}\xi = x$, $i^{*}\eta = y$, and $\alpha \in H^{k-1}(X;Q/Z)$ with $\beta(\alpha) = \xi$. Then

$$\begin{aligned} \mu(x,y) &= \langle i^{*}(\alpha) \cup y, [Y] \rangle \\ &= \langle i^{*}(\alpha) \cup i^{*}(\xi), [Y] \rangle \\ &= \langle \delta i^{*}(\alpha) \cup \xi, [X,Y] \rangle = 0 \end{aligned}$$

so $H \subset K^{\perp}$. □

Proof of V.3.1: For a lattice $L \subset V$ take $L = \{\mathrm{Im}(H^{k}(X,Y;Z) \longrightarrow H^{k}(X;Z)) \otimes Q\}$.

The dual lattice L^{+} can be pulled back to a summand of $H^{k}(X;Z)/TOR$ and $0 \longrightarrow L \longrightarrow L^{+} \longrightarrow H/K \longrightarrow 0$ is exact. Now consider the diagram

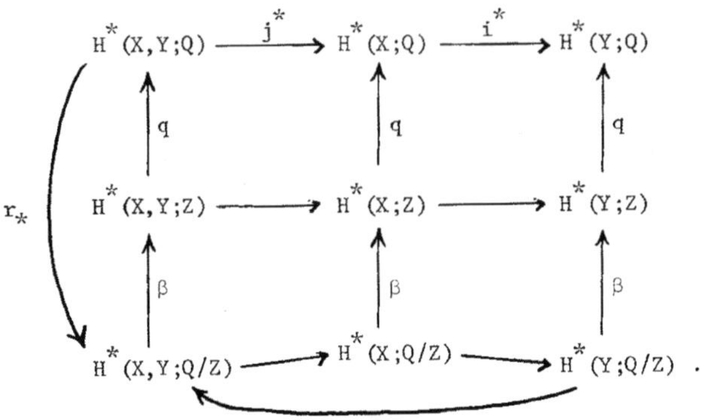

If $[\xi]$, $[\eta] \in H/K$ pick $u \in H^{k-1}(Y;Q/Z)$ with $\beta(u) = \xi$. Then the linking form on G induces the form on H/K by

$$\mu([\xi],[\eta]) = \langle u \cup \eta, [Y] \rangle \in Q/Z .$$

On the other hand, suppose $\xi = i^* x$, $\eta = i^*(y)$. We know that $i^* g(x) = 0$ because ξ is a torsion element. Let $z \in H^k(X,Y;Q)$ be such that $j^*(z) = q(x)$.

The form on $L^+/L = H/K$ is given by

$$\langle [\xi],[\eta] \rangle = \langle z \cup i^*(y), [X,Y] \rangle = \langle r^* z \cup y, [X,Y] \rangle \mod Z .$$

Now $\mu([\xi],[\eta]) = \langle u \cup \eta, [Y] \rangle = \langle \delta u \cup y, [X,Y] \rangle$. If we knew that $r^*(z) = -\delta u$ we would be finished. The fact that $r^*(z) = -\delta u$ may be established by representing the classes by cochains and chasing through the previous diagram. Therefore

$$Lk(V) = -\partial\langle \pi, V, b \rangle . \quad \square$$

(3.2) <u>Lemma</u>: <u>In</u> $H^2(C_2, GR(Z_p, C_\ell))$,

$$cl(H^s(C_{p^r}; H^m(M;Z)/TOR)) = cl(H^{m+s}(M^{2m} \times S^{2s-1}/T_0 \times \tau \circ Z))$$

<u>where</u> $s \gg m$, $T_0 = T^\ell$, $\tau(z_1, \ldots, z_s) = (\lambda z_1, \ldots, \lambda z_s)$, $\lambda = \exp(2\pi i/p^r)$, <u>and</u> C_ℓ <u>acts</u>

on $M \times S/T_0 \times \tau$ by $g[x,z] = [gx, z]$.

Proof. If I denotes the vectors in $V = H^m(M;Q)$ that are fixed by C_{p^r}, then by lemma II.1.4

$$\partial(\langle p^r\rangle \otimes I) = H^2(C_{p^r}; H^m(M;Z)/TOR) .$$

By the inductive step we can restrict our attention to $T_0^{p^{r-1}}$ being non-trivial, then M/T_0 is a rational Poincare duality space with a natural integral orientation so that the projection $M \longrightarrow M/T_0$ has degree p^r. M/T_0 admits an orientation preserving action of C_ℓ, so it determines an element $\omega(T_0,M) \in W_*(Q,C_\ell)$. Clearly $\langle p^r\rangle \otimes I = \omega(T_0,M)$ because $\mathrm{Im}\{H^m(M/T_0;Q) \longrightarrow H^m(M;Q)\} = I$.

Consider the action on the algebraic curve $R = \{[z_0,z_1,z_2] \in CP(2) \mid (z_0)^{p^r} + (z_1)^{p^r} + (z_2)^{p^r} = 0\}$ of genus $(p^r-1)(p^r-2)/2$ given by $t([z_0,z_1,z_2]) = [z_0,z_1,\lambda z_2]$. If $\varepsilon\colon Z[C_{p^r}] \longrightarrow Z$ is the augmentation homomorphism then $H^1(R;Z) = (\ker \varepsilon)^{p^r-2}$. Given a free abelian C_n-module V, tensoring over Z with the exact sequence $0 \longrightarrow \ker\varepsilon \longrightarrow Z[C_{p^r}] \longrightarrow Z \longrightarrow 0$ shows $H^i(C_{p^r};V) = H^{i+1}(C_{p^r}; V \otimes \ker\varepsilon)$. Therefore because p^n-2 is odd, in $H^2(C_2; GR(F_p, C_\ell))$, we have

$$(3.3) \quad cl(H^i(C_{p^r};H^m(M;Z)/TOR)) = (p^r-2)cl(H^i(C_{p^r};H^m(M;Z)/TOR))$$
$$= cl(H^{i+1}(C_{p^r};H^{m+1}(M \times R;Z)/TOR)) .$$

The map $t\colon R \longrightarrow R$ is a semifree action of period p^r with p^r isolated fixed points. Consider the product action on $R \times \overset{s}{\ldots} \times R = R^s$ of $t \times \overset{s}{\ldots} \times t = (\overset{s}{\underset{1}{X}} t)(\overset{s}{\underset{1}{X}} t)$ has p^{rs} isolated fixed points. Let $\{D_i \mid 1 \le i \le p^{rs}\}$ be a collection of equivariant discs about these points. If $X = R^s - \underset{i}{\cup} D_i$ then $\partial X = -\underset{i}{\cup} S_i^{2s-1}$. Notice that $H^q(M \times R^s;Z) \longrightarrow H^q(M \times X;Z)$ is an isomorphism for $q < 2s-1$.

$C_{p^r} \times C_\ell$ acts on $M \times R^s$ by letting C_ℓ act trivially on R^s. If $s > m+1$ then in $H^2(C_2; GR(F_p, C_\ell))$ we have

$$\begin{aligned}
cl(H^s(C_{p^r}; H^m(M;Z)/TOR)) &= cl(H^{2s}(C_{p^r}; H^{m+s}(M \times R^s;Z)/TOR)) && \text{by 3.3} \\
&= cl(\partial(\omega(T_0 \times (\overset{s}{\underset{1}{X}}\ t); M \times R^s))) && \text{by II.1.6} \\
&= cl(\partial(\omega(T_0 \times (\overset{s}{\underset{1}{X}}\ t), M \times X))) \\
&= p^{rs}\ cl(Lk(M \times S^{2s-1}/T_0 \times \tau)) && \text{by lemma V.3.1} \\
&= cl(H^{m+s}(M \times S^{2s-1}/T_0 \times \tau; Z)) && \text{because } p \text{ is odd.} \quad \square
\end{aligned}$$

(3.4) Lemma: Let $F^{2a} \subset M^{2m}$ be a component of the fixed set of $C_p \subset C_{p^r}$, $\tilde{C}_{p^{r-1}} \times C_\ell$ denote the quotient group which acts on F with torsion signatures $\tilde{f}$ and $\tilde{g}$, and let f and g be the torsion signatures for the action of $C_{p^r} \times C_\ell$ on M. If $r \geq 2$,

$$f = \sum_F (m-a+1)\tilde{f} + \sum_F (m-a)\tilde{g}$$

$$g = \sum_F (m-a)\tilde{f} + \sum_F (m-a+1)\tilde{g} .$$

Proof. For large s, by the Thom isomorphism theorem,

$$\begin{aligned}
H^{m+s}(M \times S^{2s-1}/T_0 \times \tau; Z) &= \bigoplus_F H^{m+s}(F \times S^{2s-1}/T_0 \times \tau; Z) \\
&= \bigoplus_F H^{a+(s+(m-a))}(F \times S^{2(s+(m-a))-1}/T_0 \times \tau; Z) .
\end{aligned}$$

Apply lemma V.3.2 to both sides

$$cl(H^s(C_{p^r}; H^m(M;Z)/TOR)) = \sum_F cl(H^{s+(m-a)}(C_{p^r}; H^a(F;Z)/TOR)) .$$

Suppose s is odd. Then the left hand side for C_{p^r} and $C_{p^{r-1}}$ gives us f.

Where $(m-a)$ is even the right hand side would be f' and when $(m-a)$ is odd the right hand side would be g' where f' and g' are the torsion signatures for the action of $C_{p^r} \times C_\ell$ on F. By Theorem V.2.1, $f' = \tilde{f}$ and $g' = \tilde{g}$. Since $2\tilde{f} = 0$ this can be expressed as

$$f = \sum_F ((m-a+1)\tilde{f} + (m-a)\tilde{g}) \ .$$

Similarly for g. □

If $n = p^r \ell$, let $F^i(T^\ell)$ be the fixed set of T^ℓ of codimension i mod 4. C_ℓ acts on $F^i(T^\ell)$ with fixed set $F^i(T^\ell) \cap F(T)$.

(3.5) Theorem:

$$f(T,p) = \chi(F^2(T^\ell) \cap F(T))\overline{cl}(F_p)$$

and $$g(T,p) = \chi(F^0(T^\ell) \cap F(T))\overline{cl}(F_p) \ .$$

Proof. The proof is by induction on r, $n = p^r \ell$. Suppose $r = 1$, let F^{2a} be a component of $F(T^\ell)$. We want to compute $\overline{cl}(H^a(F;Z)/TOR \otimes F_p)$.

$$\begin{aligned}
\overline{cl}(H^a(F;Z)/TOR \otimes F_p) &= \overline{cl}(H^a(F;F_p)) && \text{by the universal coefficient theorem} \\
&= \overline{cl}(H^*(F;F_p)) && \text{by Poincare duality} \\
&= \overline{cl}(C^*(F) \otimes F_p) && \text{by the standard Euler characteristic arguments} \\
&= \overline{cl}(C^*(F \cap F(T)) \otimes F_p && \text{because non-trivial permutation representations are 0 mod } E_\ell \\
&= \chi(F \cap F(T))\overline{cl}(F_p)
\end{aligned}$$

because $\chi(F \cap F(T)) \equiv \dim_{F_p} C^*(F \cap F(T)) \otimes F_p$ mod 2.

Therefore for s large and odd,

$$\sum_{\substack{m-a\\ odd}} \overline{cl}(H^a(F;Z)/TOR \otimes F_p) = \sum_F \overline{cl}(H^a(F)/TOR \otimes H^{s+(m-a)}(S/\tau;Z))$$

$$= \sum_F \overline{cl}(H^{m+s}(F \times (S/\tau);Z))$$

$$= \overline{cl}(H^{m+s}(M \times S/T_0 \times \tau;Z))$$

$$= \overline{cl}(H^1(C_p;H^m(M;Z)/TOR))$$

$$= f(T,p).$$

If $(m-a)$ is odd then $F \subset F^2(T^\ell)$ so that

$$f(T,p) = \chi(F^2(T^\ell) \cap F(T))\overline{cl}(F_p) .$$

Similarly $g(T,p) = \chi(F^0(T^\ell) \cap F(T))\overline{cl}(F_p)$.

Now suppose $r \geq 2$. Let $\tilde{f}$ and $\tilde{g}$ be the torsion signatures defined for $\tilde{C}_{p^{r-1}} \times C_\ell$ on F_{2a} by component of the fixed set of $C_p \subset C_{p^r}$. By lemma V.3.4 and induction

$$F(T,p) = \sum_F \{(m-a+1)\chi(\tilde{F}^2(T^\ell) \cap \tilde{F}(T))$$
$$+ (m-a)\chi(\tilde{F}^0(T^\ell) \cap F(T))\}\overline{cl}(F_p)$$

where the codimensions of $\tilde{F}^2$ and $\tilde{F}^0$ are computed in F. But this is exactly $\chi(F^2(T^\ell) \cap F(T))\overline{cl}(F_p)$ where codimensions are calculated in M. Similarly for $g(T,p)$. $\square$

Similarly $g(T,p) = \chi(F^0(T^\ell) \cap F(T))\overline{cl}(F_p)$.

(3.6) Corollary: If π is an odd order group then $\langle\pi, H^m(M^{2m};Z)/TOR\rangle \in W_*(Z,\pi)$ is uniquely determined by the Atiyah-Singer-Segal multisignature and the torsion signatures $f(T,p)$.

Proof. By Frohlich [F] We know that the natural map

$$W_*(Z,\pi) \longrightarrow W_*(\mathbb{R},\pi)$$

has the torsion subgroup of $W_*(Z,\pi)$ for its kernel. The image of $\langle\pi, H^m(M;Z)/TOR\rangle$ in $W_*(\mathbb{R},\pi)$ is uniquely determined by the Atiyah-Singer-Segal multisignature. Torsion elements, by Dress' Induction Theorem, are uniquely determined by the torsion signatures $f(T,p)$. □

In fact, since the Atiyah-Singer-Segal multisignature invariant is computed by restricting to cyclic subgroups we have

(3.7) Corollary: $\langle\pi, H^m(M;Z)/TOR\rangle$ is uniquely determined by the local data associated with the cyclic subgroups of π, when $|\pi|$ is odd. □

4. Example

In this section we describe an example of an action of C_{15} on a compact four-manifold M^4 where the Atiyah-Singer-Segal multisignature is zero but the torsion signatures are not zero. In [CF], Conner-Floyd describe an action of C_{pq} where p,q are distinct odd primes on a Riemann surface with exactly one fixed point. We will describe in detail this construction in the case of C_{2p} where p is an odd prime.

To begin consider the algebraic curve $S = \{[z_1, z_2, z_3] \mid z_1^{2p} + z_2^{2p} + z_3^{2p} = 0\}$. On S, there are two periodic maps of order $2p$ given by

$$T[z_1, z_2, z_3] = [z_1, -z_2, \lambda_p z_3]$$

and

$$\tau[z_1, z_2, z_3] = [\lambda_{2p} z_1, z_2, z_3] .$$

(4.1) Lemma: *If* $0 < k < 2p$, $T^k[z_1, z_2, z_3] = [z_1, z_2, z_3]$ *where* $[z_1, z_2, z_3] \in S$, *then either*

(i) $k \equiv 0$ *mod* 2, $z_3 = 0$ *and* $(z_1/z_2)^{2p} = -1$ *or*

(ii) $k \equiv 0$ *mod* p, $z_2 = 0$ *and* $(z_1/z_3)^{2p} = -1$.

Proof. First $z_1 \neq 0$, because if $z_1 = 0$ then both $z_2 \neq 0$ and $z_3 \neq 0$. If $T^k[z_1, z_2, z_3] = [z_1, z_2, z_3]$ then $(-1)^k z_2 = \alpha z_2$ and $(\lambda_p)^k z_3 = \alpha z_3$. This implies that $(-1)^k = \lambda_p^k$ which is not possible for $0 < k < 2p$. Assume $z_1 = 1$, then $T^k[1, z_2, z_3] = [1, z_2, z_3]$ where $(-1)^k z_2 = z_2$ and $\lambda_p^k z_3 = z_3$. This clearly happens when $z_2 = 0$ and $k \equiv 0$ mod p or $z_3 = 0$ and $k \equiv 0$ mod 2. □

(4.2) Lemma: *If* $0 < k < 2p$ *and* $k \equiv 0$ *mod* 2 *then the fixed point set of* $T^k: S \longrightarrow S$ *is* $\{[1, \pm i\lambda_p^j, 0] \mid 0 \leq j < p\}$ *while if* $k \equiv 0$ *mod* p *then the fixed point set of* T^k *is* $\{[1, 0, \pm i\lambda_p^j] \mid 0 \leq j < p\}$.

Proof. Follows easily from V.4.1. □

(4.3) Lemma: *If* $0 < k < 2p$, *then* τ *and* T^k *have no coincidences.*

Proof. Suppose $T^k[z_1, z_2, z_3] = \tau[z_1, z_2, z_3]$, then $z_1 \neq 0$ because if $z_1 = 0$

$$[0, (-1)^k z_2, \lambda_p^k z_3] = [0, z_2, z_3]$$

which violates V.4.1. Let $z_1 = 1$, then

$$[1,(-1)^k z_2,\lambda_p^k z_3] = [\lambda_{2p},z_2,z_3] = [1,\lambda_{2p}^{-1}z_2,\lambda_{2p}^{-1}z_3] .$$

Therefore if $z_2 \neq 0$, then $(-1)^k = \lambda_{2p}^{-1}$ while if $z_3 \neq 0$, then $(\lambda_p)^k = \lambda_{2p}^{-1}$ both of which are impossible. □

Let $M = S/T$ and define $\tau^*(\langle s\rangle) = \langle \tau(s)\rangle$. This is well-defined and is a map of period $2p$ on M with exactly one fixed point; the one covered by $\{[0,\pm i\lambda_p^j,1] \mid 0 \leq j < p\}$. Let $\tau_0 = (\tau^*)^2 : M \longrightarrow M$. τ_0 has two additional fixed points a and b covered by $A = \{[1,0,i\,\lambda_p^j] \mid 0 \leq j < p\}$ and $B = \{[1,0,-i\lambda_p^j] \mid 0 \leq j < p\}$ respectively. (That is $a = \langle 1,0,i\rangle$ and $b = \langle 1,0,-i\rangle$ where $[z_1,z_2,z_3] \in S$ represents the equivalence class $\langle z_1,z_2,z_3\rangle \in M$.)

Let us analyze the local data for τ_0 around the fixed points $x_0 = \langle 0,i,1\rangle$, a, and b. Consider the homogeneous coordinate patch $P_3 = \{[z_1,z_2,1] \mid z_1,z_2 \in C\}$. $S \cap P_3$ is the Riemann surface for the function $(1+\xi^{2p})^{1/2p}$ embedded as follows $\xi \longrightarrow [\xi, i(1+\xi^{2p})^{1/2p},1] \in S$. This is a multivalued function: as long as $(1+\xi^{2p}) \neq 0$ we have $2p$ distinct roots to choose from. This means $S \cap P_3$ is actually a 2p-fold branched covering of C branched at $(2p)$-th roots of -1, that is, where $1+\xi^{2p} = 0$.

Now we must describe the way T acts on $S \cap P_3$. $T([z_1,z_2,1]) = [z_1,-z_2,\lambda_p] = [\lambda_p^{-1}z_1,(-\lambda_p^{-1})z_2,1]$ hence T cyclically permutes the branch points. The set $\{[0,\pm i\lambda_p^j,1] \mid 0 \leq j < p\}$ that covers the fixed point x_0 in M can be indexed $\xi_j = [0,i(-\lambda_p^{-1})^j,1]$ where $0 \leq j < 2p$. Then $T(\xi_j) = \xi_{j+1}$, $j < 2p-1$, and $T(\xi_{2p-1}) = \xi_0$. This set is the fiber above $0 \in C$.

Consider $D = \{\xi \in C \mid |\xi| < 1/2\}$. On D it is possible to find a well-defined branch of the function $(1+\xi^{2p})^{1/2p}$. Call this function $f(\xi)$. There are now 2p-embeddings of D into $S \cap P_3$ by $f_j(\xi) = [\xi, i(-\lambda_p^{-1})^j f(\xi),1]$. Call this image disc D_j. Since

$$\begin{aligned} T[\xi,i(-\lambda_p^{-1})^j f(\xi),1] &= [\lambda_p^{-1}\xi, i(-\lambda_p^{-1})^{j+1} f(\xi),1] \\ &= [\lambda_p^{-1}\xi, i(-\lambda_p^{-1})^{j+1} f(\lambda_p^{-1}\xi),1] , \end{aligned}$$

we see that $T: D_j \longrightarrow D_{j+1}$ with a rotation of λ_p^{-1}.

τ acts on these discs as follows. Since

$$\tau[\xi, i(-\lambda_p^{-1})^j f(\xi), 1] = [\lambda_{2p}\xi, i(-\lambda_p^{-1})^j f(\xi), 1]$$
$$= [\lambda_{2p}\xi, i(-\lambda_p^{-1})^j f(\lambda_{2p}\xi), 1]$$

we see that $\tau: D_j \longrightarrow D_j$ with a rotation of $2\pi/2p$ radians.

If $M = S/T$, then $D_0/T \approx D_0$ is a neighborhood of the fixed point x_0 and τ^* acts on the normal bundle by multiplication by λ_{2p}. Therefore the local contribution of x_0 in the Atiyah-Singer-Segal multisignature is $(\lambda_{2p}+1)/(\lambda_{2p}-1)$.

(4.4) <u>Lemma</u>: $\operatorname{sgn}(M, \tau^*) = (\lambda_{2p}+1)/(\lambda_{2p}-1)$ <u>and for</u> $n = pq$, p, q <u>distinct odd primes</u>

$$\operatorname{sgn}(M, \tau^*) = (\lambda_{pq}+1)/(\lambda_{pq}-1) .$$

Proof. The reader is referred to [AS]. □

Now consider the homogeneous coordinate patch $P_1 = \{[1, z_2, z_3] \mid z_2, z_3 \in C\}$. Again $S \cap P_1$ is the Riemann surface for the function $(1+\xi^{2p})^{1/2p}$ with the embedding being given by

$$\xi \longrightarrow [1, \xi, i(1+\xi^{2p})^{1/2p} .$$

This is a $2p$ branched covering of C, branched at ξ a $2p$-th root of -1. The fixed point a of τ_0 is covered by $A = \{[1, 0, i\lambda_p^j] \mid 0 \le j < p\}$ and b is covered by $B = \{[1, 0, -i\lambda_p^j] \mid 0 \le j < p\}$. $A \cup B$ is fiber over $0 \in C$.

Index the elements in A by $\alpha_j = [1, 0, i\lambda_p^j]$ and those in B by $\beta_j = [1, 0, -i\lambda_p^j]$. Then $T(\alpha_j) = \alpha_{j+1}$, $T(\alpha_{p-1}) = \alpha_0$ and $T(\beta_j) = \beta_{j+1}$ and $T(\beta_{p-1}) = \beta_0$. Again let $D = \{\xi \in C \mid |\xi| \le \frac{1}{2}\}$. Then there are $2p$ embedding of D into $S \cap P_1$ by

$$f_j(\xi) = [1, \xi, i\lambda_p^j(1+\xi^{2p})^{1/2p}], \quad 0 \le j < p$$

and $$g_j(\xi) = [1, \xi, -i\lambda_p^j(1+\xi^{2p})^{1/2p}], \quad 0 \le j < p .$$

Let $D_j^+ = f_j(D)$ and $D_j^- = g_j(D)$. Since

$$T[1,\xi, \pm i\lambda_p^j(1+\xi^{2p})^{1/2p}] = [1, -\xi, \pm i\lambda_p^{j+1}(1+(-\xi)^{2p})^{1/2p}].$$

T maps $D_i^{\pm}$ to $D_{i+1}^{\pm}$ followed by a rotation of π-radians. In particular T^p: $D_j^{\pm}$ $D_j^{\pm}$ by a rotation through π-radians. Now

$$(4.5) \quad \tau^2[1,\xi,\pm i\lambda_p^j(1+\xi^{2p})^{1/2p}] = [\lambda_{2p}^2, \xi, \pm i\lambda_p^j(1+\xi^{2p})^{1/2p}]$$
$$= [1, \lambda_p^{-1}\xi, \pm i\lambda_p^{j-1}(1+(\lambda_p^{-1}\xi)^{2p})^{1/2p}]$$
$$\sim [1, -\lambda_p^{-1}\xi, \pm i\lambda_p^{j}(1+(\lambda_p^{-1}\xi)^{2p})^{1/2p}]$$
$$\sim [1, \lambda_p^{-1}\xi, \pm i\lambda_p^{j}(1+(\lambda_p^{-1}\xi)^{2p})^{1/2p}].$$

The map from $D_0^{\pm} \longrightarrow D_0^{\pm}/T$ is a double branched covering and τ_0 acting on $D_0^{\pm}/T$ is covered by τ^2 acting on $D_0^{\pm}$ as described by formula V.4.5. That is $\tau_0(\langle\xi\rangle) = \langle\tau^2\xi\rangle = \langle\lambda_p^{-1}\xi\rangle = \lambda_p^{-2}\langle\xi\rangle$ because of the double covering. Therefore the local data of τ_0 at a and b is the same and τ_0 acts in the normal bundle by multiplication by λ_p^{-2}.

(4.6) __Lemma__: $\operatorname{sgn}(M,\tau_0) = \dfrac{\lambda_p+1}{\lambda_p-1} - 2\,\dfrac{\lambda_p^2+1}{\lambda_p^2-1}$. __More generally for__ $n = pq$

$$\operatorname{sgn}(M, (\tau^*)p) = \frac{\lambda_q+1}{\lambda_q-1} - p\,\frac{\lambda_q^p+1}{\lambda_q^p-1} .$$

Proof. The first term is the local contribution to the Atiyah-Singer-Segal multisignature from x. The second term comes from the fixed points a and b. The general formula is entirely similar. □

For convenience we collect some formulas for $\lambda = \lambda_{15}$.

$$\begin{aligned}
\lambda+\lambda^{-1}+\lambda^{2}+\lambda^{-2}+\lambda^{4}+\lambda^{-4}+\lambda^{7}+\lambda^{-7} &= 1\\
\lambda^{-1}-\lambda+\lambda^{-2}-\lambda^{2}+\lambda^{-4}-\lambda^{4}+\lambda^{-7}-\lambda^{7} &= (\lambda+1)/(\lambda-1)\\
\lambda^{3}+\lambda^{6}+\lambda^{9}+\lambda^{12} &= -1\\
\lambda^{5}+\lambda^{10} &= -1 .
\end{aligned}$$

Let M_p denote the Riemann surface constructed for $2p$ and let $\tau_p = \tau_0$ and M_{pq} denote the surface for pq, p,q odd and $\tau_{pq} = \tau$. Now consider $(\tau_3 \times \tau_5, M_3 \times M_5)$.

(4.7) Lemma: If $(T,M) = (\tau_3 \times \tau_5, M_3 \times M_5)$, then

$$\begin{aligned}
\operatorname{sgn}(T,M) &= (\lambda+\lambda^{-1}) - (\lambda^{2}+\lambda^{-2}) - (\lambda^{4}+\lambda^{-4}) + (\lambda^{7}+\lambda^{-7})\\
\operatorname{sgn}(T^{3},M) &= 0\\
\operatorname{sgn}(T^{5},M) &= 0\\
\operatorname{sgn}(T^{15},M) &= 0 .
\end{aligned}$$

Proof. $\operatorname{sgn}(T,M) = \operatorname{sgn}(\tau_3,M_3)\operatorname{sgn}(\tau_5,M_5)$

$$= \left(\frac{\lambda^5+1}{\lambda^5-1} - 2\,\frac{\lambda^{10}+1}{\lambda^{10}-1}\right)\left(\frac{\lambda^3+1}{\lambda^3-1} - 2\,\frac{\lambda^6+1}{\lambda^6-1}\right)$$

where $\lambda = \lambda_{15}$ by V.4.6. So that

$$\begin{aligned}
\operatorname{sgn}(T,M) &= (\lambda^{-5}-\lambda^{5})(-\lambda^{3}+\lambda^{-3}+\lambda^{6}-\lambda^{-6})\\
&= (\lambda+\lambda^{-1}) - (\lambda^{2}+\lambda^{-2}) - (\lambda^{4}+\lambda^{-4}) + (\lambda^{7}+\lambda^{-7}) .
\end{aligned}$$

The other calculations follow because for a trivial action

$\operatorname{sgn}(T',M') = \operatorname{sgn}(M')$ and $\operatorname{sgn}(M') = 0$ for a surface. □

Let $N = M_3 \times M_{15}$ and $T_N = \tau_3 \times \tau_{15}$.

(4.8) Lemma:

$$\mathrm{sgn}(T_N,N) = -(\lambda+\lambda^{-1}) - (\lambda^2+\lambda^{-2}) - (\lambda^4+\lambda^{-4}) + (\lambda^7+\lambda^{-7}) + 2(\lambda^6+\lambda^{-6})$$

$$\mathrm{sgn}(T_N^3,N) = 0$$

$$\mathrm{sgn}(T_N^5,N) = 6$$

$$\mathrm{sgn}(T_N^{15},N) = 0 \ .$$

Proof. $\mathrm{sgn}(T_N,N) = \mathrm{sgn}(\tau_3,M_3)\mathrm{sgn}(\tau_{15},M_{15})$

$$= (\lambda^{-5}-\lambda^5)(\lambda^{-1}-\lambda+\lambda^{-2}-\lambda^2+\lambda^{-4}-\lambda^4+\lambda^{-7}-\lambda^7)$$

$$= (\lambda+\lambda^{-1}) - (\lambda^2+\lambda^{-2}) - (\lambda^4+\lambda^{-4}) + (\lambda^7+\lambda^{-7}) + 2(\lambda^6+\lambda^{-6})$$

$\mathrm{sgn}(T_N^5,N) = \mathrm{sgn}(\tau_3^2,M_3)\mathrm{sgn}(\tau_{15}^5,M_{15}) = (\lambda^5-\lambda^{-5})\cdot\left(\frac{\lambda^5+1}{\lambda^5-1} - 5\,\frac{\lambda^{10}+1}{\lambda^{10}-1}\right)$ by V.4.6.

$= 2|\lambda^5-\lambda^{-5}|^2 = 6.$ □

Let $\mathcal{R}_k = C_{15}/C_k \times CP(2)$ where $k = 1,3,5,$ or 15. C_{15} acts on $\mathcal{R}_k$ by coset multiplication on C_{15}/C_k. Call this map ρ_k. We collect the information in the last few lemmas in a chart.

	(T_M,M)	(T_N,N)	$(\rho_1,\mathcal{R}_1)$	$(\rho_2,\mathcal{R}\ )$	$(\rho_5,\mathcal{R}_5)$	$(\rho_{15},\mathcal{R}_{15})$
T	$(\lambda+\lambda^{-1})$ $-(\lambda^2+\lambda^{-2})$ $-(\lambda^4+\lambda^{-4})$ $+(\lambda^7+\lambda^{-7})$	$-(\lambda+\lambda^{-1})$ $-(\lambda^2+\lambda^{-2})$ $-(\lambda^4+\lambda^{-4})+2(\lambda^6+\lambda^{-6})$ $+(\lambda^7+\lambda^{-7})$	0	0	0	1
T^3	0	0	0	0	3	1
T^5	0	6	0	5	0	1
T^{15}	0	0	15	5	3	1

Consider $M' = M \cup N \cup N \cup \mathcal{R}_1 \cup -3\mathcal{R}_3 \cup -\mathcal{R}_5 \cup 3\mathcal{R}_{15}$ and the map $T'\colon M' \quad M'$ by

$$T' = T_M \cup T_N^4 \cup T_N^8 \cup \rho_1 \cup 3\rho_3 \cup \rho_5 \cup 3\rho_{15} \ .$$

The reader should show that

$$\operatorname{sgn}(T',M') = 0, \quad \operatorname{sgn}(T'^3,M') = 0, \quad \operatorname{sgn}(T'^5,M') = 0,$$

and $\operatorname{sgn}(T'^{15},M') = 0$. Hence the Atiyah-Singer-Segal multisignature vanishes on (T',M').

Now consider the torsion signature

$$f(T',3) = \chi(F^2((T')^5) \cap F(T'))\overline{\mathrm{cl}}(F_3) .$$

Since f is additive we compute f for each component of M'. On M, the fixed set of T_M^5 is $\mathrm{Fix}(\tau_3) \times M_5$ and the fixed set of T_M is $\mathrm{Fix}(\tau_3) \times \mathrm{Fix}(\tau_5)$ which is 9 points. Hence $\chi(F^2(T_M^5) \cap F(T_M)) = 9$. Because $5 \equiv -1 \bmod 3$ we have that $\overline{\mathrm{cl}}(F_3) \neq 0 \in H^2(C_2; GR(F_3, C_5))/E_3$ so that $f(T_M, 3) = \overline{\mathrm{cl}}(F_3) \neq 0$. For T_N, $\mathrm{Fix}(T_N^5) = \mathrm{Fix}(\tau_3) \times \mathrm{Fix}(\tau_{15}^5)$ which is a disjoint union of points. $F^2(T_N^5) = \emptyset$ so $f(T_N^4, 3) = f(T_N^8, 3) = 0$.

Similarly $f(\rho_1, 3) = 0$, $f(\rho_3, 3) = 0$, $f(\rho_5, 3) = 0$ and $f(\rho_{15}, 3) = 0$ because the codimension 2 mod 4 component of the fixed set is always empty. Hence $f(T',3) = f(T_M, 3) = \mathrm{cl}(F_3) \neq 0$.

In fact, $\mathrm{TOR}(W_0(Z, C_{15})) = Z_2$ and the integral form $(C_{15}, H^2(M', Z))$ represents the torsion element.

Chapter VI
Metabolic and Hyperbolic Forms

In this chapter we present two approaches to the metabolic-hyperbolic problem. The first covers the general case and can be applied to groups of odd prime order. In the second approach we restrict our attention to cyclic groups of odd prime power order and eventually study the natural homomorphism $L_0^h(C_{p^n}) \longrightarrow W_0(Z, C_{p^n})$. The first section defines the functor $\mathrm{Met}_\pi(N)$ associated to metabolic π-modules V with $N \subset V$ so that $N = N^\perp$. The group $\mathrm{Met}_\pi(N)$ is related to $\mathrm{Ext}^1_{Z(\pi)}(N^*, N)$ in a natural way and Theorem 1.2 shows that $Met_\pi(N)$ injects into $Ext^1_{Z(\pi)}(N^*,N)$. In section 2 further functorial properties of $Met_\pi(N)$ are developed and Theorem 1.2 is extended to Theorem 2.2 to show that the image of $Met_\pi(N)$ into $Ext^1_{Z(\pi)}(N^*,N)$ is exactly the subgroup of elements fixed by dualizing an extension. Furthermore a homomorphism $Ext^1_{Z(\pi)}(N^*,N) \longrightarrow Met_\pi(N)$ is defined and its kernel computed. The major result in this direction is Theorem 3.2. This result reduces the study of whether a metabolic module is hyperbolic to indecomposable π-representations. In the case of a cyclic group of (odd) prime order these indecomposable representations are well known and Theorem 4.3 proves that a metabolic integral orthogonal representation of type II is hyperbolic.

Section five begins an alternate analysis of when a metabolic module of type II is hyperbolic. We restrict our attention to π-modules V that are π-projective and by lemma 4.1 we only need to find a projective metabolizer. This is accomplished for cyclic groups of prime power order. The next several sections provide the information to compute the kernel and cokernel of

$$L_0^h(C_{p^k}) \longrightarrow W_0(Z, C_{p^k})$$

for p an odd prime. This recovers in a special case results of Wall, Bak, Pardon, and Karoubi.

1. $Met_\pi(N)$

With π a finite group of odd order we fix an integral representation (π,N). Now consider all $Z(\pi)$-module embeddings

$$0 \longrightarrow (\pi,N) \xrightarrow{\rho} (\pi,V,b)$$

where

i) (π,V,b) is an integral orthogonal representation of type II

ii) $im(\rho)^\perp = im(\rho)$.

We shall say that $\rho\colon (\pi,N) \longrightarrow (\pi,V,b)$ is Baer equivalent to $\sigma\colon (\pi,N) \longrightarrow (\pi,W,b_1)$ if and only if there is a π-module isometry J for which

$$\begin{array}{ccc} & \nearrow & (\pi,V,b) \\ (\pi,N) & & \simeq \;\downarrow J \\ & \searrow & (\pi,W,b_1) \end{array}$$

commutes. These Baer classes form an abelian group called $\mathrm{Met}_\pi(N)$ where the additive identity will be the hyperbolic form $H(\pi,N)$. That is, $N^* = \mathrm{Hom}_Z(N,Z)$ and $(g\varphi)(x) \equiv \varphi(g^{-1}x)$ is the dual of (π,N) and on $(\pi,N) \oplus (\pi,N^*)$ the hyperbolic innerproduct is $h((x,\varphi),(y,\psi)) = \psi(x) + \varphi(y)$. Then $\rho\colon (\pi,N) \longrightarrow H((\pi,N))$ is the inclusion $x \longrightarrow (x,0)$.

We describe addition first. Given $\rho\colon (\pi,N) \longrightarrow (\pi,V,b)$ and $\sigma\colon (\pi,N) \longrightarrow (\pi_1,W,b_1)$ we form the orthogonal sum $(\pi, V\oplus W, b\oplus b_1)$ and take $L \subset V \oplus W$ to be the π-submodule $(\rho(x),-\sigma(x))$ for all $x \in N$. Now $L \subset L^\perp$ and $L^\perp$ is the submodule of all pairs (v,w) with $b(v,\rho(x)) = b_1(w,\sigma(x))$ for all $x \in N$. There is induced a $(\pi, L^\perp/L, \beta)$. If $\langle v,w\rangle$ denotes the element in $L^\perp/L$ corresponding to $(v,w) \in L^\perp$ then

$$\beta(\langle v,w\rangle, \langle v_1,w_1\rangle) = b(v,v_1) + b_1(w,w_1) \ .$$

This is an integral orthogonal representation of type II. We take

$$\rho + \sigma\colon (\pi,N) \longrightarrow L^\perp/L$$

to be $x \quad \langle\rho(x),0\rangle = \langle 0,\sigma(x)\rangle$. We should prove that the image of $(\rho+\sigma)$ is equal to its own orthogonal complement. Suppose $\langle v,w\rangle$ is in this orthogonal complement. Then for all $x \in N$

$$b(v,\rho(x)) + b_1(w,0) = b(v,0) + b_1(w,\sigma(x)) = 0 \ .$$

This says $v \in \mathrm{im}(\rho)^\perp = \mathrm{im}(\rho)$ and $w \in \mathrm{im}(\sigma)^\perp = \mathrm{im}(\sigma)$. Write $\rho(x_1) = v$, $\sigma(x_2) = w$, then

$$\langle v,w\rangle = \langle \rho(x_1),\sigma(x_2)\rangle = \langle \rho(x_1) + \rho(x_2),\sigma(x_2) - \sigma(x_2)\rangle$$
$$= \langle \rho(x_1 + x_2),0\rangle$$

as required. Thus the addition operation is defined in $\mathrm{Met}_\pi(N)$.

The reader should check that $H((\pi,N))$ is the additive identity and that $(\pi,N) \xrightarrow{-\rho} (\pi,V,-b)$ is the additive inverse of $(\pi,N) \xrightarrow{\rho} (\pi,V,b)$.

The group $\mathrm{Met}_\pi(N)$ is related to $\mathrm{Ext}^1_{Z(\pi)}(N^*,N)$ as follows. For $0 \to (\pi,N) \xrightarrow{\rho} (\pi,V)$ we define $\bar{\rho}: V \to N^*$ by $\bar{\rho}(v)(x) = b(v,\rho(x))$ and this yields a short exact sequence of $Z(\pi)$-modules

$$0 \to (\pi,N) \xrightarrow{\rho} (\pi,V) \xrightarrow{\bar{\rho}} (\pi,N^*) \to 0$$

which in turn defines an element in $\mathrm{Ext}^1_{Z(\pi)}(N^*,N)$.

(1.1) Lemma: The map $\mathrm{Met}_\pi(N) \to \mathrm{Ext}^1_{Z(\pi)}(N^*,N)$ given by $((\pi,N) \to (\pi,V,b)) \to (0 \to (\pi,N) \xrightarrow{\rho} (\pi,N) \xrightarrow{\bar{\rho}} (\pi,N^*) \to 0)$ is a homomorphism.

Proof. From $0 \to (\pi,N) \xrightarrow{\rho} (\pi,V,b)$ and $0 \to (\pi,N) \xrightarrow{\rho_1} (\pi,W,b_1)$ we have

$$0 \to N \xrightarrow{\rho} V \xrightarrow{\bar{\rho}} N^* \to 0$$
$$0 \to N \xrightarrow{\rho_1} W \xrightarrow{\bar{\rho}_1} N^* \to 0 .$$

With $L = \{(-\rho(x),\rho_1(x)) \mid x \in N\}$, then $L^\perp \subset V \oplus W$ is exactly the submodule of (v,w) with $\bar{\rho}(v) = \bar{\rho}_1(w)$. Thus we have

$$0 \to N \oplus N \xrightarrow{\rho\oplus\sigma} L^\perp - N^* \to 0 .$$

To define addition in $\mathrm{Ext}_{Z(\pi)}(N^*,N)$ we would form $N \oplus L^\perp$ and factor out the subgroup $\Gamma = (-x-y,\rho(x),\rho_1(y))$. We must show $N \oplus L^\perp/\Gamma$ is suitably equivalent to $L^\perp/L$. With $z \in N$, $(v,w) \in L^\perp$ we send $N + L^\perp \to L^\perp/L$ by

$$(z,(v,w)) \longrightarrow \langle v+\rho(z), w\rangle \ .$$

Note Γ is the kernel. First $(-x-y, \rho(x), \rho_1(y)) \longrightarrow \langle -\rho(y), \rho_1(y)\rangle = 0$. Then if $\langle v+\rho(z), w\rangle = 0$ there is a $y \in N$ with $-\rho(y) = v+\rho(z)$ and $\rho_1(y) = w$. Thus $(a,(v,w)) = (y+z-y, \rho(-y-z), \rho_1(y)) \in \Gamma$. Also $(z,(0,0)) \longrightarrow \langle \rho(z), 0\rangle$. Thus our definition of addition in $\mathrm{Met}_\pi(N)$ agrees with addition in $\mathrm{Ext}^1_{Z(\pi)}(N^*, N)$ and we have a homomorphism. □

(1.2) <u>Theorem</u>: <u>The homomorphism</u> $\mathrm{Met}_\pi(N) \longrightarrow \mathrm{Ext}^1_{Z(\pi)}(N^*, N)$ <u>is a monomorphism</u>.

Proof. This will be rather involved and uses the fact that all (π, V, b) are of type II.

Let us assume that there is a $Z(\pi)$-module section of $0 \longrightarrow N \xrightarrow{\rho} V \xrightarrow{\bar{\rho}} N^* \longrightarrow 0$. This is, $f: N^* \longrightarrow V$ is a $Z(\pi)$-module homomorphism with $\bar{\rho} f = \text{identity}$. Using this we have an isomorphism

$$N \oplus N^* \xrightarrow{\rho \oplus f} V \ .$$

Then b on V will induce an innerproduct on $N \oplus N^*$. Let us see what this is like.

Write $v = \rho(x) + f(\varphi)$ and $w = \rho(y) + f(\psi)$. Then

$$\begin{aligned} b(v,w) &= b(\rho(x),\rho(y)) + b(\rho(x), f(\psi)) \\ &\quad + b(\rho(y), f(\varphi)) + b(f(\varphi), f(\psi)) \ . \end{aligned}$$

But $b(\rho(x),\rho(y)) = 0$ and $b(\rho(x), f(\psi)) = (\bar{\rho} f)(\psi)(x) = \psi(x), b(\rho(y), f(\varphi)) = \varphi(y)$. Thus on $N \oplus N^*$ the innerproduct induced by b is

$$\langle (x,\varphi), (y,\psi)\rangle = \varphi(y) + \psi(x) + \beta(\varphi,\psi) \ .$$

Here $\beta(\varphi,\psi)$ is some Z-valued bilinear form on N^*. Of course β is symmetric, of type II and $\beta(g\varphi, g\psi) = \beta(\varphi,\psi)$.

Recalling the canonical isomorphism $(\pi,N) \simeq (\pi,N^{**})$ we may define $s\colon N^* \longrightarrow N$ so that

$$\psi(s(\varphi)) = \beta(\varphi,\psi).$$

Thus $s \in \mathrm{Hom}_Z(\pi)(N^*,N)$. But then there is the adjoint of s; namely, $s^* \in \mathrm{Hom}_{Z(\pi)}(N^*,N)$ defined so that $\varphi(s^*(\psi)) = \psi(s(\varphi)) = \beta(\varphi,\psi)$. Since $\beta(\varphi,\psi) = \beta(\psi,\varphi)$ we have $s = s^*$.

Suppose now there is $e \in \mathrm{Hom}_{Z(\pi)}(N^*,N)$ with $e + e^* = s$. We can alter the cross-section $f\colon (\pi,N^*) \longrightarrow (\pi,V)$ into

$$f'(\varphi) = f(\varphi) - \rho(e(\varphi)) .$$

Of course $\bar{\rho}f' = \text{identity}$. We claim $N \oplus N^* \xrightarrow{\rho \oplus f'} V$ is a Baer equivalence of $0 \longrightarrow (\pi,N) \xrightarrow{\rho} (\pi,V,b)$ with the hyperbolic form. We simply write $v = \rho(x) + f'(\varphi)$ and $w = \rho(y) + f'(\psi)$ so that

$$\begin{aligned} b(v,w) &= \psi(x) + \varphi(y) + b(f'(\varphi),f'(\psi)) \\ &= \psi(x) + \varphi(y) + \beta(\varphi,\psi) - b(f(\varphi),\rho(e(\psi))) \\ &\quad - b(\rho(e(\varphi)),f(\psi)) \\ &= \psi(x) + \varphi(y) + \psi(s(\varphi)) - \varphi(e(\psi)) - \psi(e(\varphi)) \\ &= \psi(x) + \varphi(y) + (s(\varphi)) - \psi(e^*(\varphi)) - \psi(e(\varphi)) \\ &= \psi(x) + \varphi(y) . \end{aligned}$$

Thus if we think of $h \longrightarrow h^*$ as an action of C_2 on $\mathrm{Hom}_{Z(\pi)}(N^*,N)$ we want to prove $\mathrm{cl}(s) = 0 \in H^2(C_2,\mathrm{Hom}_{Z(\pi)}(N^*,N))$.

Let us look first at $\mathrm{Hom}_Z(N^*,N)$. This is a π-module with

$$(g \circ h)(\varphi) \equiv gh(g^{-1}\varphi)$$

for $g \in \pi$, $h \in \mathrm{Hom}_Z(N^*,N)$, $\varphi \in N^*$. Clearly $\mathrm{Hom}_{Z(\pi)}(N^*,N) \subset \mathrm{Hom}_Z(N^*,N)$ as the subgroup of π-fixed homomorphisms.

The action of C_2 on $\mathrm{Hom}_Z(N^*,N)$ is to send $h \longrightarrow h^*$ where

$$\psi(h(\varphi)) = \varphi(h^*(\psi)) .$$

We want to show the action of π commutes with this involution; that is, $(g \circ h)^* = g \circ (h^*)$. Now

$$\varphi((g \circ h)^*(\psi)) = \psi(g \circ h(\varphi)) = (g^{-1}\psi)(h(g^{-1}\varphi))$$
$$= g^{-1}\varphi(h^*(g^{-1}\psi)) = \varphi(g \circ h^*(\psi)).$$

Thus we have the required $(g \circ h)^* = g \circ (h^*)$.

Next we define Σ: $\mathrm{Hom}_Z(N^*,N) \longrightarrow \mathrm{Hom}_{Z(\pi)}(N^*,N)$ by

$$\Sigma(h) = \sum_{g \in \pi} g \circ h .$$

Then Σ is a C_2-module homomorphism and the composition

$$0 \quad \mathrm{Hom}_{Z(\pi)}(N^*,N) \xrightarrow{i} \mathrm{Hom}_Z(N^*,N) \xrightarrow{\Sigma} \mathrm{Hom}_{Z(\pi)}(N^*,N)$$

is multiplication by the group order which is odd. This proves

(1.3) Lemma: The inclusion $\mathrm{Hom}_{Z(\pi)}(N^*,N) \rightarrow \mathrm{Hom}_Z(N^*,N)$ induces a monomorphism $0 \rightarrow H^2(C_2;\mathrm{Hom}_{Z(\pi)}(N^*,N)) \rightarrow H^2(C_2;\mathrm{Hom}_Z(N^*,N))$. □

It is this latter group which is easily computed. Choose a basis $e_1,\ldots,e_n$ for N together with a dual basis $e_1^*,\ldots,e_n^*$ in N^*. Then $h \in \mathrm{Hom}_Z(N^*,N)$ is represented by the integral matrix

$$h(e_i^*) = \sum_j A_{ji} e_j$$

and h^* is represented by the transpose $[A_{ij}]$. Thus $h = h^*$ if and only if

$[A_{ij}]$ is symmetric and $h = e + e^*$ if and only if every diagonal entry in A is even. This last is equivalent to $\beta(\varphi,\psi) = \varphi(h(\psi))$ being of type II.

With (VI.1.3) and the comment about the need for showing $cl(s) = 0$ in $H^2(C_2;\mathrm{Hom}_{Z(\pi)}(N^*,N))$ we have completed the proof of (VI.1.2). □

(1.4) Corollary: Let (π,V,b) be an integral orthogonal representation of type II. Then the diagonal embedding

$$d\colon (\pi,V) \longrightarrow (\pi, V \oplus V,\ b \oplus -b)$$

represents $0 \in \mathrm{Met}_\pi(V)$.

Proof. We need only show $0 \longrightarrow V \xrightarrow{d} V \oplus V \xrightarrow{D} V^* \longrightarrow 0$ splits. For any $\varphi\colon V \longrightarrow Z$ there is a unique $v_\varphi \in V$ with $b(v_\varphi, w) \equiv \varphi(w)$. Now $f(\varphi) = (v_\varphi, 0)$ is the required $Z(\pi)$-module splitting. □

2. Functorial properties

We can regard $\mathrm{Met}_\pi(N)$ as a covariant functor on the category of integral representations and $Z(\pi)$-module homomorphisms.

Suppose $k\colon (\pi,N) \longrightarrow (\pi,N_1)$ is a $Z(\pi)$-module homomorphism and we are given $d(N \xrightarrow{\rho} V) \in \mathrm{Met}_\pi(N)$. Form $V \oplus (N_1 \oplus N_1^*)$ and take L to be the set of all triples $(\rho(x), -k(x), 0)$. Now $L^\perp = \{(v,y,\varphi) \mid b(v,\rho(x)) = \varphi(k(x)) = (k^*\varphi)(x)\}$. Define

$$k_\#(\rho)\colon N_1 \longrightarrow L^\perp/L$$

by

$$z \longrightarrow \langle 0, z, 0\rangle .$$

If $\langle v,y,\varphi\rangle$ is in the complement of $\mathrm{im}(k_\#(\rho))$ then $b(v,0) + \varphi(z) = 0$ and hence $\varphi = 0$. But then $b(v,\rho(x)) \equiv 0$ and $v \in \mathrm{im}(\rho)^\perp = \mathrm{im}(\rho)$ so we write $v = \rho(x_1)$. Now $\langle v,y,\varphi\rangle = \langle \rho(x_1),y,0\rangle = \langle \rho(x_1)-\rho(x_1), y+k(x_1), 0\rangle = \langle 0, y+k(x_1), 0\rangle$. Thus $\mathrm{im}(k_\#(\rho))^\perp = \mathrm{im}(k_\#(\rho))$. In this manner

$$k_*(\mathrm{cl}(N \xrightarrow{\rho} V)) = \mathrm{cl}(N_1 \xrightarrow{k_\#(\rho)} L^\perp/L)$$

defines

$$k_*\colon \mathrm{Met}_\pi(N) \longrightarrow \mathrm{Met}_\pi(N_1) \ .$$

Note $k\colon (\pi, N) \longrightarrow (\pi, N_1)$ induces

$$k_*^1\colon \mathrm{Ext}^1_{Z(\pi)}(N^*, N) \longrightarrow \mathrm{Ext}^1_{Z(\pi)}(N^*, N_1)$$

and that

$$k^*\colon (\pi, N_1^*) \longrightarrow (\pi, N^*)$$

induces

$$k_*^2\colon \mathrm{Ext}^1_{Z(\pi)}(N^*, N_1) \longrightarrow \mathrm{Ext}^1_{Z(\pi)}(N_1^*, N_1)$$

so that if $k_*\colon \mathrm{Ext}^1_{Z(\pi)}(N^*, N) \longrightarrow \mathrm{Ext}^1_{Z(\pi)}(N_1^*, N_1^*)$ is $k_*^2 \circ k_*^1$ then

$$\begin{array}{ccccc} 0 \longrightarrow & \mathrm{Met}_\pi(N) & \longrightarrow & \mathrm{Ext}^1_{Z(\pi)}(N^*, N) \\ & \downarrow k_* & & \downarrow k_* \\ 0 \longrightarrow & \mathrm{Met}_\pi(N_1) & \longrightarrow & \mathrm{Ext}^1_{Z(\pi)}(N_1^*, N_1^*) \end{array}$$

commutes.

If $k\colon (\pi, N) \longrightarrow (\pi, N_1) \longrightarrow 0$ is an epimorphism then an alternative description of $k_*\colon \mathrm{Met}_\pi(N) \longrightarrow \mathrm{Met}_\pi(N_1)$ can be given. For $0 \longrightarrow (\pi, N) \longrightarrow (\pi, V, b)$ let $\mathcal{L} \subset V$ be the image under ρ of the kernel of k. Then $\mathcal{L}$ is a Z-module summand of V and $\mathcal{L} \subset \mathcal{L}^\perp$. We define $K(\rho)\colon N_1 \longrightarrow \mathcal{L}^\perp/\mathcal{L}$ so that

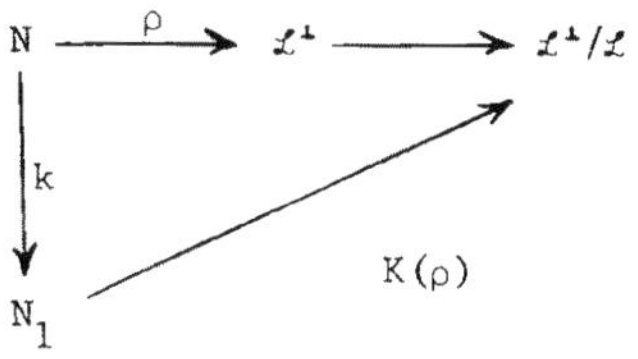

commutes.

(2.1) <u>Lemma</u>: <u>The</u> $N_1 \xrightarrow{N(\rho)} \mathcal{L}^\perp/\mathcal{L}$ <u>represents</u> $k_*(cl(N \xrightarrow{\rho} V))$ <u>in</u> $Met_\pi(N_1)$.

Proof. Recall the $L \subset V \oplus N_1 \oplus N_1^*$ and the description of $L^\perp$. If $v \in \mathcal{L}^\perp$ then v determines a homomorphism $\varphi_v \in N_1^*$ by $\varphi_v(x_1) = b(v,\rho(x)) = \bar{\rho}(v)(x)$ for any x with $k(x) = x_1$. Thus $\langle v,0,\varphi_v\rangle \in L^\perp/L$. If $\langle v,0,\varphi_v\rangle = 0 \in L^\perp/L$ then for some $x \in N$, $(\rho(x),-k(x),0) = (v,0,\varphi_v)$ but then $k(x) = 0$ and $x \in \ker(k)$ so $v \in L$. Thus $\mathcal{L}^\perp/\mathcal{L} \longrightarrow L^\perp/L$ is a monomorphism.

Next given any $\langle v,y,\varphi\rangle \in L^\perp/L$ then $b(v,\rho(x)) = \varphi(k(x))$ so $b(v,\rho(x)) = 0$ if $x \in \ker(k)$ and hence $v \in \mathcal{L}^\perp$ and $\varphi = \varphi_v$. There is $y_0 \in N$ with $k(y_0) = y$ so $\langle v,y,\varphi_v\rangle = \langle v+\rho(y_0),0,\varphi_v\rangle$. Now $v+\rho(y_0) \in \mathcal{L}^\perp$ also and $\varphi_v = \varphi_v + \rho(y_0)$. Thus we find $\mathcal{L}^\perp/\mathcal{L} \longrightarrow L^\perp/L$ is an isomorphism. Finally $\beta(\langle v,0,\varphi_v\rangle,\langle w,0,\varphi_w\rangle) = b(v,w)$ and so we have an isometry.

The $k_\#(\rho)\colon N_1 \longrightarrow L^\perp/L$ was $z \longrightarrow \langle 0,z,0\rangle$. Since $k(z_0) = z$ has a solution, $\langle 0,z,0\rangle = \langle\rho(z_0),0,0\rangle$ which is the image of $K(\rho)(z)$ under the isometry $\mathcal{L}^\perp/\mathcal{L} \longrightarrow L^\perp/L$. □

There is a natural involution $\mathcal{D}$ on $Ext^1_{Z(\pi)}(N^*,N)$. If

$$0 \longrightarrow (\pi,N) \xrightarrow{\alpha} (\pi,K) \xrightarrow{\beta} (\pi,N^*) \longrightarrow 0$$

is short exact then so is the dual sequence, $\beta^*(x)(k) = \beta(k)(x), \alpha^*(\varphi)(x) = \varphi(\alpha(x))$,

$$0 \longrightarrow (\pi,N) \xrightarrow{\beta^*} (\pi,K^*) \xrightarrow{\alpha^*} (\pi,N^*) \longrightarrow 0$$

where we use the canonical isomorphism $(\pi, N) \simeq (\pi, N^{**})$. This defines the involution on $\operatorname{Ext}^1_{Z(\pi)}(N^*, N)$.

Not surprisingly every element in the image of $0 \longrightarrow \operatorname{Met}_\pi(N) \longrightarrow \operatorname{Ext}^1_{Z(\pi)}(N^*, N)$ is fixed under this involution. If $0 \longrightarrow (\pi, N) \xrightarrow{\ \rho\ } (\pi, V, b)$ is given then the Baer equivalence

$$\begin{array}{ccccccc} & & & V^* & & & \\ & & \overset{\vec{\rho}^{\,*}}{\nearrow} & \uparrow J & \overset{\rho^*}{\searrow} & & \\ 0 \longrightarrow & N & & & & N^* & \longrightarrow 0 \\ & & \underset{\rho}{\searrow} & & \underset{\bar{\rho}}{\nearrow} & & \\ & & & V & & & \end{array}$$

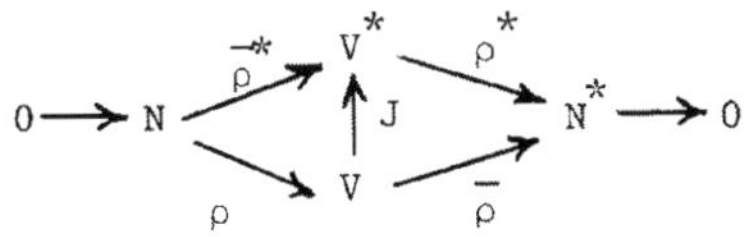

is given by $J(v)(w) \equiv b(v, w)$. Note $\rho^*(J(v))(x) = J(v)(\rho(x)) = b(v, \rho(x)) = \bar{\rho}(v)(x)$ so $\bar{\rho}(v) = \rho^*(J(v))$. Also $\vec{\rho}^{\,*}(x)(v) = \bar{\rho}(v)(x) = b(v, \rho(x)) = J(\rho(x))(v)$ and thus $\vec{\rho}^{\,*}(x) = J(\rho(x))$.

We shall now introduce a homomorphism $\operatorname{Ext}^1_{Z(\pi)}(N^*, N) \longrightarrow \operatorname{Met}_\pi(N)$. This will be done in two steps. Begin with

$$0 \longrightarrow (\pi, N) \xrightarrow{\ \alpha\ } K \xrightarrow{\ \beta\ } (\pi, N^*) \longrightarrow 0$$

and pass to

$$(\pi, N \oplus N) \xrightarrow{\ (\alpha, \beta^*)\ } (\pi, K \oplus K^*) .$$

The reader should check that with respect to the hyperbolic form on $K \oplus K^*$ the image of (α, β^*) is its own orthogonal complement.

So we have thus far a homomorphism $\operatorname{Ext}^1_{Z(\pi)}(N^*, N) \longrightarrow \operatorname{Met}_\pi(N \oplus N)$. The homomorphism $N \oplus N \longrightarrow N \longrightarrow 0$ given by $(x, y) \longrightarrow x + y$ induces $\operatorname{Met}_\pi(N \oplus N) \longrightarrow \operatorname{Met}_\pi(N)$. Using the composition we arrive at

$$\operatorname{Ext}^1_{Z(\pi)}(N^*, N) \longrightarrow \operatorname{Met}_\pi(N) .$$

Let $\mathcal{D}\colon \operatorname{Ext}^1_{Z(\pi)}(N^*, N) \longrightarrow \operatorname{Ext}^1_{Z(\pi)}(N^*, N)$ be duality.

The composition

$$0 \longrightarrow \mathrm{Met}_{\pi}(N) \longrightarrow \mathrm{Ext}^1_{Z(\pi)}(N^*, N) \longrightarrow \mathrm{Met}_{\pi}(N)$$

is multiplication by 2. The composition $\mathrm{Ext}^1_{Z(\pi)}(N^*,N) \longrightarrow \mathrm{Met}_{\pi}(N) \longrightarrow \mathrm{Ext}^1_{Z(\pi)}(N^*,N)$ is the operator $(\mathcal{D}+I)$. Since $\mathrm{Ext}^1_{Z(\pi)}(N^*,N) \approx H^1(\pi;\mathrm{Hom}_Z(N^*,N))$, the group $\mathrm{Ext}^1_{Z(\pi)}(N^*,N)$ is a finite abelian group of odd order so any $\mathcal{D}$-fixed element lies in the image of $(\mathcal{D}+I)$.

(2.2) Theorem: <u>The image of $0 \longrightarrow \mathrm{Met}_{\pi}(N) \longrightarrow \mathrm{Ext}^1_{Z(\pi)}(N^*,N)$ is the subgroup of $\mathcal{D}$-fixed elements. The kernel of $\mathrm{Ext}^1_{Z(\pi)}(N^*,N) \longrightarrow \mathrm{Met}_{\pi}(N) \longrightarrow 0$ is the image of the operator</u> $(\mathcal{D}-I)$.

Proof. Since $\mathrm{Met}_{\pi}(N)$ has odd order multiplication by 2 is an automorphism. Also $\mathrm{Ext}^1_{Z(\pi)}(N^*,N) \approx \mathrm{Im}(\mathcal{D}+I) \oplus \mathrm{Im}(\mathcal{D}-I)$. □

3. Direct sums

We have the situation that $0 \longrightarrow (\pi,N) \xrightarrow{\ \rho\ } (\pi,V,b)$ might be non-zero in $\mathrm{Met}_{\pi}(N)$ while there is another $0 \longrightarrow (\pi,L) \xrightarrow{\ \sigma\ } (\pi,V,b)$ which is $0 \in \mathrm{Met}_{\pi}(L)$. This means (π,V,b) is hyperbolic, but not on $\mathrm{im}(\rho)$. In this section we shall see an example of this phenomenon.

Consider the direct sum of two integral representations $N_1 \oplus N_2$. There are the projection homomorphisms $t^i\colon N_1 \oplus N_2 \longrightarrow N_i \longrightarrow 0$.

(3.1) Lemma: <u>There is induced an epimorphism</u>

$$(t^1_*, t^2_*)\colon\ \mathrm{Met}_{\pi}(N_1 \oplus N_2) \longrightarrow \mathrm{Met}_{\pi}(N_1) \oplus \mathrm{Met}_{\pi}(N_2)\ .$$

Proof. Given $0 \longrightarrow (\pi,N_1) \longrightarrow (\pi,V,b)$ we form $(\pi,N_1 \oplus N_2) \xrightarrow{\ r\ } (\pi, V \oplus N_2 \oplus N_2^*, b+h)$ where $r(x_1,x_2) = (\rho(x_1),x_2,0)$. This is an element in $\mathrm{Met}_{\pi}(N_1 \oplus N_2)$. Since t^i is an epimorphism we can use (VI.2.1) to study t^i_* on this element. For t^1 the kernel is $(0,N_2)$ and hence $\mathcal{L} = (0,N_2,0)$. The complement $\mathcal{L}^{\perp}$ is $V \oplus N_2 \oplus \{0\}$. Thus $\mathcal{L}^{\perp}/\mathcal{L} = V$ and clearly

$$t_*^1(N_1 \oplus N_2 \longrightarrow V \oplus N_2 \oplus N_2^*) = (N_1 \longrightarrow V) .$$

Now for t^2 the kernel is N_1 and now $\mathcal{L}$ is $(\rho(x),0,0)$. The complement $\mathcal{L}^{\perp}$ is $\rho(N_1) \oplus (N_2 \oplus N_2^*)$ and $\mathcal{L}^{\perp}/\mathcal{L}$ is the hyperbolic $H(N_2)$ so t_*^2 takes $(N_1 \oplus N_2 \longrightarrow V \oplus N_2 \oplus N_2^*)$ into $0 \in \mathrm{Met}_\pi(N_2)$. Reversing the roles of N_1 and N_2 completes the proof of (VI.3.1). □

We want to understand the kernel of the above homomorphism. To do this we need a homomorphism

$$\mathrm{Ext}^1_{Z(\pi)}(N_2^*, N_1) \longrightarrow \mathrm{Met}_\pi(N_1 \oplus N_2) .$$

Briefly from

$$0 \longrightarrow N_1 \xrightarrow{\alpha} K \xrightarrow{\beta} N_2^* \longrightarrow 0$$

we use $0 \longrightarrow (N_1 \oplus N_2) \xrightarrow{(\alpha,\beta^*)} K \oplus K^*$ with respect to the hyperbolic form on $K \oplus K^*$ to define the element in $\mathrm{Met}_\pi(N_1 \oplus N_2)$. The point to be carefully noted is that while $N_1 \oplus N_2 \xrightarrow{(\alpha,\beta^*)} K \oplus K^*$ definitely need not be 0 in $\mathrm{Met}_\pi(N_1 \oplus N_2)$ it is clear that $K \oplus K^*$ is hyperbolic.

(3.2) Theorem: The sequence

$$0 \longrightarrow \mathrm{Ext}^1_{Z(\pi)}(N_2^*, N_1) \longrightarrow \mathrm{Met}_\pi(N_1 \oplus N_2) \xrightarrow{(t_*^1, t_*^2)} \mathrm{Met}_\pi(N_1) \oplus \mathrm{Met}_\pi(N_2) \longrightarrow 0$$

is split short exact.

Proof. To see $0 \longrightarrow \mathrm{Ext}_{Z(\pi)}(N_2^*, N_1) \longrightarrow \mathrm{Met}_\pi(N_1 \oplus N_2)$ we note $(N_1 \oplus N_2)^* \simeq N_1^* \oplus N_2^*$ so that

$$\begin{aligned}&\mathrm{Ext}^1_{Z(\pi)}((N_1 \oplus N_2)^*, N_1 \oplus N_2)\\ &\quad \simeq \mathrm{Ext}^1_{Z(\pi)}(N_1^*, N_1) \oplus \mathrm{Ext}^1_{Z(\pi)}(N_2^*, N_2)\\ &\quad \oplus (\mathrm{Ext}^1_{Z(\pi)}(N_2^*, N_1) \oplus \mathrm{Ext}^1_{Z(\pi)}(N_1^*, N_2)) .\end{aligned}$$

The effect of the duality operator is to interchange $\mathrm{Ext}^1_{Z(\pi)}(N_2^*, N_1)$ and $\mathrm{Ext}^1_{Z(\pi)}(N_1^*, N_2)$. Then if $\gamma \in \mathrm{Ext}^1_{Z(\pi)}(N_1^*, N_2)$ the composition

$$\mathrm{Ext}^1_{Z(\pi)}(N_2^*, N_1) \longrightarrow \mathrm{Met}_\pi(N_1 \oplus N_2) \longrightarrow \mathrm{Ext}^1_{Z(\pi)}(N_1^* \oplus N_2^*, N_1 \oplus N_2)$$

takes γ to $\gamma + \mathcal{D}(\gamma)$. This last is 0 if and only if $\gamma = 0$.

We would like to use an explicit computation to show exactness at the middle term. Suppose $0 \longrightarrow (\pi, N_1 \oplus N_2) \xrightarrow{\rho} (\pi, V, b)$ lies in the kernel of both t^1_* and t^2_*. Let $\mathcal{L}_1 \subset V$ denote $\rho(N_2)$, which is the image under ρ of the kernel of t^1. Let $\rho_1 \colon N_1 \longrightarrow \mathcal{L}_1^\perp/\mathcal{L}_1$ be induced by $\rho\colon N_1 \longrightarrow \mathcal{L}^\perp$. Since $(\pi, N_1) \xrightarrow{\rho_1} \mathcal{L}_1^\perp/\mathcal{L}_1 = 0 \in \mathrm{Met}_\pi(N_1)$ there is a π-invariant submodule $\mathcal{N}_1 \subset \mathcal{L}_1^\perp/\mathcal{L}$ for which

$$\begin{aligned}\mathrm{im}(\rho_1) \cap \mathcal{N}_1 &= \{0\}\\ \mathrm{im}(\rho_1) + \mathcal{N}_1 &= \mathcal{L}_1^\perp/\mathcal{L}_1\\ \mathcal{N}_1^\perp &= \mathcal{N}_1 .\end{aligned}$$

We can identify $\mathcal{N}_1$ with N_1^* by restricting $\bar{\rho}_1\colon \mathcal{L}_1^\perp/\mathcal{L}_1 \longrightarrow N_1^*$ down to $\mathcal{N}_1$.

Let $S_1 \subset \mathcal{L}_1^\perp \subset V$ be the inverse image of $\mathcal{N}_1$ with respect to $\mathcal{L}_1^\perp \longrightarrow \mathcal{L}_1^\perp/\mathcal{L}_1$. Then S_1 is π-invariant and $S_1 \cap \mathrm{im}(\rho) = \rho(N_2) = \mathcal{L}_1$. If $v \in S_1^\perp$ then $v \in \mathcal{L}_1^\perp$ and since $\mathcal{N}_1^\perp = \mathcal{N}_1$ we can find $w \in \mathcal{L}_1$ for which $v-w \in S_1$. This shows $S_1^\perp = S_1$. We repeat this with $\mathcal{L}_2 = \rho(N_1)$ and $\mathcal{N}_2 \subset \mathcal{L}_2^\perp/\mathcal{L}_2$ to obtain $S_2 \subset \mathcal{L}_2^\perp$ which is also π-invariant and for which $S_2^\perp = S_2$, $S_2 \cap \mathrm{im}(\rho) = \rho(N_1) = \mathcal{L}_2$.

Now $\mathcal{L}_1^\perp \cap \mathcal{L}_2^\perp = \mathrm{im}(\rho)$ since $\mathcal{L}_1 + \mathcal{L}_2 = \mathrm{im}(\rho) = \mathrm{im}(\rho)^\perp$. However $S_1 \cap S_2 \subset \mathcal{L}_1^\perp \cap \mathcal{L}_2^\perp = \mathrm{im}(\rho)$ and $(S_1 \cap S_2) \cap \mathrm{im}(\rho) = \{0\}$ so $S_1 \cap S_2 = \{0\}$. Next we assert $S_1 + S_2 = V$. If

$v \in V$ then v induces $\varphi_v: N_1 \longrightarrow Z$ by $\varphi_v(x_1) = b(v,\rho(x_1))$. On the other hand $\bar{\rho}_1/\mathcal{N}_1 \simeq N_1^*$ so there is a $w_1 \in S_1 \subset \mathcal{L}_1^{\perp}$ with $v-w_1 \in \rho(N_1)^{\perp} = \mathcal{L}_2^{\perp}$. We can then similarly find $w_2 \in S_2 \subset \mathcal{L}_2^{\perp}$ with $v-w_1-w_2 \in \rho(N_2)^{\perp} = \mathcal{L}_1^{\perp}$. In fact $v-w_1-w_2 \in \mathcal{L}_1^{\perp} \cap \mathcal{L}_2^{\perp} = \mathrm{im}(\rho) = S_1 \cap \mathrm{im}(\rho) + S_2 \cap \mathrm{im}(\rho)$. Thus $S_1 + S_2 = V$.

Now we have

$$0 \longrightarrow N_1 \xrightarrow{\ \rho\ } S_2 \xrightarrow{\ \bar{\rho}_2\ } N_2^* \longrightarrow 0$$

and

$$0 \longrightarrow N_2 \xrightarrow{\ \rho_2\ } S_1 \xrightarrow{\ \bar{\rho}_1\ } N_1^* \longrightarrow 0 .$$

We wish to identify the second sequence as the dual of the first. If $\varphi: S_2 \quad Z$ is a homomorphism in S_2^* then there is $v \in V$ with $\beta(v,s) = \varphi(s)$ for all $s \in S_2$. We write $v = s_1 + s_2$, $s_i \in S_i$ so $\beta(s_1,s) = \varphi(s)$. This $s_1 \in S_1$ is unique so we have $J: S_1 \simeq S_2^*$. We must show

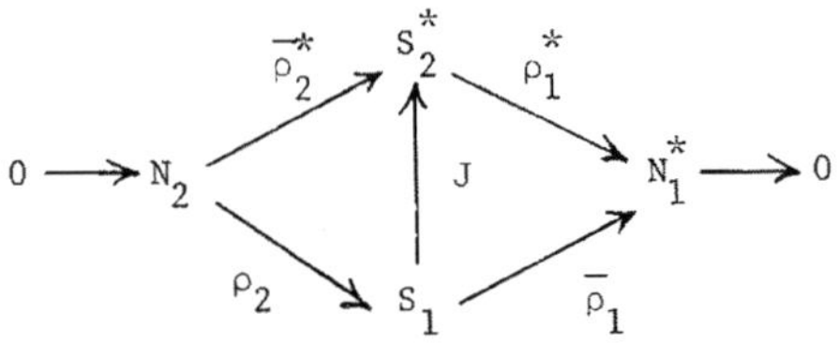

commutes. For $s \in S_2$, $\bar{\rho}_2(s)(y) = b(s,\rho_2(y))$ for all $y \in N_2$. Therefore $\bar{\rho}_2^*(y)(s) = b(\rho_2(y),s) = J(\rho_2(y))$. Similarly $\rho_2 \circ J = \rho_1$.

Thus our kernel element is the image of $0 \longrightarrow N_1 \xrightarrow{\ \rho_1\ } S_2 \xrightarrow{\ \bar{\rho}_2\ } N_2^* \longrightarrow 0$ under $\mathrm{Ext}^1_\pi(N_2^*, N_1) \longrightarrow \mathrm{Met}_\pi(N_1 \oplus N_2)$.

To split the sequence use $i^1: N_1 \longrightarrow N_1 \oplus N_2$, $i^1(x_1) = (x_1, 0)$ and $i^2: N_2 \longrightarrow N_1 \oplus N_2$, $i^2(x_2) = (0, x_2)$. Then $(t_*^1 i_*^1, t_*^2 i_*^2) = \mathrm{id}, \mathrm{id})$. □

4. Hyperbolic forms

An orthogonal integral representation (π, V, b) is hyperbolic if and only if it is isometrically equivalent to the hyperbolic form on some integral representation. Now we can easily say

(4.1) *Lemma: If an integral orthogonal representation of type* II *admits a metabolizer which is projective as a* $Z(\pi)$*-module then the orthogonal representation is hyperbolic.*

Proof. By assumption there is a $0 \longrightarrow (\pi,P) \xrightarrow{\rho} (\pi,V,b)$ with $\operatorname{im}(\rho)^{\perp} = \operatorname{im}(\rho)$ and P a $Z(\pi)$-projective. But then $\operatorname{Ext}^1_{Z(\pi)}(P^*,P) = 0$ so $\operatorname{Met}_\pi(P) = 0$. $\square$

To proceed further we must recall some definitions. An integral representation (π,N) is *decomposable* if and only if it contains a proper non-trivial $Z(\pi)$-module summand. Otherwise the representation is *indecomposable*. We shall now restrict attention to C_p a cyclic group of odd prime order. We can state that the indecomposable integral representations of C_p are known by the Reiner-Dietrichsen Theorem ([Cu-R]).

(4.2) *Lemma: If* $\mathscr{P}$ *is a finitely generated* $Z(C_p)$*-projective module and* P *is a finitely generated* $Z(\Lambda_p)$*-projective module then* $\operatorname{Met}_{C_p}(\mathscr{P}\oplus P) = 0$.

Proof. We need only show $\operatorname{Ext}^1_{Z(C_p)}(\mathscr{P}^*\oplus P^*, \mathscr{P}\oplus P) = 0$. Clearly $\operatorname{Ext}_{Z(C_p)}(\mathscr{P}^*,\mathscr{P}) = 0$ since $\mathscr{P}^*$ is also $Z(C_p)$-projective. For the same reason $\operatorname{Ext}^1_{Z(C_p)}(\mathscr{P}^*,P) = 0$. Now $\operatorname{Ext}^1_{Z(C_p)}(P^*,P) = \operatorname{Ext}^1_{Z(\Lambda_p)}(P^*,P) = 0$. Finally by duality $\operatorname{Ext}^1_{Z(C_p)}(P^*,\mathscr{P}) \simeq \operatorname{Ext}^1_{Z(C_p)}(\mathscr{P}^*,P) = 0$. $\square$

In general an integral (C_p,N) can be written as $T \oplus \mathscr{P} \oplus P$ where T is trivial representation. Now $\operatorname{Ext}^1_{Z(C_p)}(T^*,T) = 0 = \operatorname{Met}_{C_p}(T)$ surely. Appealing to (VI.3.2) and (VI.4.2) we find with $N_1 = (\mathscr{P}\oplus P)$, $N_2 = T$ that

$$\operatorname{Ext}^1_{Z(C_p)}(T^*,\mathscr{P}\oplus P) \simeq \operatorname{Met}_{C_p}(N)\ .$$

(4.3) *Theorem: If* (C_p,V,b) *is a metabolic integral orthogonal representation of type* II *then* (C_p,V,b) *is hyperbolic.* $\square$

A curious example arises by $N = Z\oplus Z(\Lambda)$. In this case $\operatorname{Ext}^1_{Z(C_p)}(Z,Z(\Lambda)) \approx$

$H^1(C_p;Z(\Lambda)) \simeq Z/p$. Using

$$0 \longrightarrow Z(\Lambda) \longrightarrow Z(C_p) \xrightarrow{\varepsilon} Z \longrightarrow 0$$

we construct

$$\rho: Z(\Lambda) \oplus Z \longrightarrow Z(C_p) \oplus Z(C_p)$$

to demonstrate that while V may be $Z(C_p)$-free we cannot assume every metabolizer is $Z(C_p)$-projective.

For a finite group of odd order we can introduce a refinement, $WH_0(Z,\pi)$, of $W_0(Z,\pi)$. We say (π,V,b) is stably hyperbolic if and only if for some integral representation (π,L) the orthogonal sum $V \oplus H(L)$ is hyperbolic. Now restricting attention only to integral orthogonal representations of type II we say (π,V,b) is stably equivalent to (π,V_1,b_1) if and only if $(\pi,V\oplus V_1, b\oplus(-b_1))$ is stably hyperbolic. Using (VI.1.3) it follows that this is an equivalence relation. The classes $[\pi,V,b]$ make up the group $WH_0(Z,\pi)$. Again by (VI.1.3) $-[\pi,V,b] = [\pi,V,-b]$.

(4.5) <u>Corollary</u>: <u>The natural homomorphism</u> $WH_0(Z,C_0) \longrightarrow W_0(Z,C_p)$ <u>is a monomorphism</u>. □

Stoltzfus [Sf] has produced elements in the kernel of $WH_0(Z,C_{p^2}) \longrightarrow W_0(Z,C_{p^2})$ as a part of his general study of the relationship of metabolic to hyperbolic.

In the next section we shall use another approach to this metabolic-hyperbolic question when (π,V,b) is restricted to be a projective $Z(\pi)$-module.

(4.6) Exercise: Develop the corresponding program for skew-symmetric (π,V,b).

5. Metabolic, even forms on projective modules

In this section we prove

(5.1) Theorem: If V is an even, metabolic, Z-nonsingular, C_{p^n} invariant innerproduct space and V is a projective $Z(C_{p^n})$-module, then V is hyperbolic.

The result rests heavily on several theorems about cohomology of groups.

(5.2) Theorem [Ca F]: If G is a p-group and A is a G-module without torsion, then the following are equivalent:

i) A is cohomologically trivial

ii) A/pA is a free $F_p[G]$-module

iii) A is a projective G-module. □

By lemma VI.4.1 we need to construct $X \subset V$ such that $X = X^\perp$ and X is a projective C_{p^n}-module. The proof is by induction on n. For $n = 0$, this is the observation that a type II innerproduct space over Z which is metabolic is also hyperbolic.

We keep our usual notation: C_p is the subgroup of order p in C_{p^n} and the quotient group is denoted $\tilde{C}_{p^{n-1}}$. C_p is generated by $t^{p^{n-1}}$ and $\Delta = 1 - t^{p^{n-1}}$ and $\Sigma = 1 + t^{p^{n-1}} + \ldots + t^{p^n - p^{n-1}}$.

Consider V a projective C_{p^n}-module with an even, C_{p^n}-invariant, Z-nonsingular, symmetric, metabolic inner product. Let $U = \{v \in V \mid \Sigma v = 0\}$. U is a Z-summand of V and it is also a projective $Z(\Lambda_{p^n})$-module. Since V is metabolic, $U \otimes Z(1/p)$ is also and hence there is a summand $W \subset U$ invariant under C_{p^n} such that $W = W^\perp \cap U$ where $W^\perp = \{v \in V \mid (v,w) = 0 \text{ for all } w \in V\}$.

(5.3) Lemma: $W^\perp/W$ is a projective $Z[\tilde{C}_{p^{n-1}}]$-module.

Proof. This follows in several steps. First we show that $t^{p^{n-1}}$ acts trivially on $W^\perp/W$. Take $w \in W^\perp$, then $\Delta w \in W^\perp$ and $\Delta w \in \ker \Sigma = U$ so that $\Delta w \in W^\perp \cap U = W$. Hence $\Delta(w+W) = \Delta w + W = 0 + W$. This makes $W^\perp/W$ a $Z[\tilde{C}_{p^{n-1}}]$-module in a natural way. To show $W^\perp/W$ is a projective $\tilde{C}_{p^{n-1}}$-module we will verify that $(W^\perp/W) \otimes F_p$ is a free $F_p(\tilde{C}_{p^{n-1}})$-module. To do this we must compute several cohomology groups.

(5.4) Lemma: $H^2(C_p;W) = 0$

$$H^1(C_p;W) = \{F_p(\tilde{C}_{p^{n-1}})\}^r.$$

Proof. Because W is a projective $Z(\lambda_{p^n})$-module and $Z(\lambda_{p^n})/(1-\lambda_p) \approx F_p(\tilde{C}_{p^{n-1}})$ we get that $H^2(C_p;W) = 0$. If $r = \dim_{Q(\lambda_{p^n})} W \otimes Q$, then $H^1(C_p;W) = W/\Delta W = [Z(\lambda_{p^n})/(1-\lambda_p)]^r = [F_p(\tilde{C}_{p^{n-1}})]^r$. □

(5.5) Lemma: $H^1(C_p;W^{\perp}) = 0$

$$H^2(C_p;W^{\perp}) = \{F_p(\tilde{C}_{p^{n-1}})\}^r.$$

Proof. The sequence $0 \to W^{\perp} \to V \to V/W^{\perp} \approx W^* \to 0$ is exact. By VI.5.4 and the fact that V is a projective C_{p^n}-module, $H^1(C_p;W^{\perp}) \approx H^2(C_p;W^*) \approx H^2(C_p;W) = 0$. Similarly for $H^2(C_p;W^{\perp})$. □

To finish the proof of lemma VI.5.3, consider the six-term cohomology sequence for $0 \to W \to W^{\perp} \to W^{\perp}/W \to 0$. It reduces to

$$0 \to H^2(C_p;W^{\perp}) \to H^2(C_p;W^{\perp}/W) \to H^1(C_p;W) \to 0 .$$

This is actually an exact sequence of $F_p(\tilde{C}_{p^{n-1}})$-modules.

Since the first and third terms are free modules we conclude that $H^2(C_p;W^{\perp}/W) = [F_p(\tilde{C}_{p^{n-1}})]^{2r}$. Now $C_p \subset C_{p^n}$ acts trivially on $W^{\perp}/W$, so

$$H^2(C_p;W^{\perp}/W) \approx (W^{\perp}/W) \otimes F_p .$$

Thus, by Theorem VI.5.2.ii, shows $W^{\perp}/W$ is a projective $Z(\tilde{C}_{p^{n-1}})$-module. □

We apply our induction hypothesis to $W^{\perp}/W$. It is a projective, $Z(\tilde{C}_{p^{n-1}})$ module with a symmetric, even, metabolic, $\tilde{C}_{p^{n-1}}$-invariant, Z-nonsingular form.

Therefore it is hyperbolic. Let $H(A)$ be the hyperbolic form on the $\tilde{C}_{p^{n-1}}$-module A. Then $W^{\perp}/W \approx H(A)$ for some A. (Note: A must be a projective $Z(\tilde{C}_{p^{n-1}})$-modnle.)

(5.6) To finish the proof we must find $B \subset W^{\perp}/W$ such that the following are true.

i) $W^{\perp}/W = H(B)$ (i.e. $B = B^{\perp}$ and $W^{\perp}/W = B \oplus B^{*}$)

ii) If $Y = \{w \in W^{\perp} | w + W \in B\}$ then
$0 \to W \to Y \to B \to 0$ is exact and $Y = Y^{\perp}$.

iii) $\delta: H^2(C_{p^n};B) \to H^1(C_{p^n};W)$ is an isomorphism.

Since $H^1(C_{p^n},B) = H^2(C_{p^n};W) = 0$, this means $H^1(C_{p^n};Y) = H^2(C_{p^n};Y) = 0$. Again by Theorem VI.5.2, Y is a projective C_{p^n}-module and the theorem is proved.

To construct B, consider the exact cohomology sequence derived from $0 \to W \to W^{\perp} \to W^{\perp}/W \to 0$, that is

$$0 \to H^2(C_{p^n};W^{\perp}) \to H^2(C_{p^n};W^{\perp}/W) \to H^1(C_{p^n};W) \to 0$$

and observe that

i) $H^2(C_{p^n};W^{\perp}) = F_p^r \approx H^1(C_{p^n};W)$

$H^2(C_{p^n};Z_p) \approx F_p^r$

ii) On $H^2(C_{p^n};W^{\perp}/W)$ there is a nonsingular Q/Z-valued innerproduct with values in $F_p \subset Q/Z$ induced by the innerproduct on $W^{\perp}/W$.

This innerproduct is metabolic (see II.1.2) and if $C = \mathrm{Im}\{H^2(C_{p^n};W^{\perp}) \to H^2(C_{p^n};W^{\perp}/W)\}$ then $C = C^{*}$. Everything is clear except $C = C^{\perp}$. By ranks we only need $C \subset C^{\perp}$. Let ΣV denote the image of Σ on V. ΣV annihilates U hence it annihilates W. Therefore $\Sigma V \subset W^{\perp}$. $W^{\perp}/\Sigma V$ is a projective $Z(\Lambda_{p^n})$-module, which implies that $H^2(C_{p^n};\Sigma V) \to H^2(C_{p^n};W^{\perp})$ is onto. Therefore we can pick representatives for innerproducts in C to come from ΣV. If $\Sigma v, \Sigma w \in C$ then

$$\langle \Sigma v, \Sigma w\rangle = p\langle v, \Sigma w\rangle \equiv 0 \bmod p .$$

So C annihilates itself.

There is another self-annihilating subspace in $H^2(C_{p^n}; W^\perp/W)$. It is

$$D = \operatorname{Im}\{H^2(C_{p^n}; A) \longrightarrow H^2(C_{p^n}; W^\perp/W)\}$$

by lemma II.1.2. If it were true that $C \cap D = \{0\}$ then the composition $H^2(C_{p^n}; A) \longrightarrow H^2(C_{p^n}; W^\perp/W) \longrightarrow H^1(C_{p^n}; W)$ is an isomorphism. Take $B = A$ in VI.5.6 to finish the proof.

Suppose $C \cap D \neq \{0\}$. We need

(5.7) Lemma: *If M is a projective $Z[\tilde{C}_{p^{n-1}}]$ module and $P \subset H^2(C_{p^n}; M)$ is an F_p summand then there is a projective summand $M_p \subset M$ such that* $\operatorname{Im}\{H^2(C_{p^n}; M_p) \longrightarrow H^2(C_{p^n}; M)\} = P$.

Proof. Since M is a projective $Z[\tilde{C}_{p^{n-1}}]$-module $H^2(C_p; M)$ is a free $F_p(\tilde{C}_{p^{n-1}})$-module. The proof is by induction on $\dim_{F_p} P$. It suffices to consider the case where $\dim_{F_p} P = 1$. Let $x_1, \ldots, x_r$ be a $F_p(\tilde{C}_{p^{n-1}})$ basis for $H^2(C_p; M)$. Then $H^2(C_{p^n}; M)$ has a basis $\tilde{\Sigma}x_1, \tilde{\Sigma}x_2, \ldots, \tilde{\Sigma}x_r$. If $x \neq 0$, $x \in P$ then $x = a_1\tilde{\Sigma}x_1 + \ldots + a_r\tilde{\Sigma}x_r = \tilde{\Sigma}(a_1x_1 + \ldots + a_rx_r)$ where some $a_i \neq 0$. But $\tilde{\xi} = a_1x_1 + \ldots + a_rx_r$ is in a basis for $H^2(C_p; M)$, so that P can be lifted back to a free $F_p(\tilde{C}_{p^{n-1}})$-module $\tilde{P} \subset H^2(C_p; M)$.

Take $\xi \in M$ so that $\xi + pM = \tilde{\xi}$. The $Z(\tilde{C}_{p^{n-1}})$ module generated by ξ is projective because $\xi Z(\tilde{C}_{p^{n-1}})/p\xi Z(\tilde{C}_{p^{n-1}}) \approx \tilde{P}$. Now let $M_p = \{x \in M \mid kx \in \xi Z(\tilde{C}_{p^{n-1}})$ for some $k \in Z\}$. As a Z-module, $\operatorname{rank}_Z M_p = \operatorname{rank}_Z \xi \cdot Z(\tilde{C}_{p^{n-1}})$ and therefore $\operatorname{rank}_{F_p} M_p/pM_p = \operatorname{rank}_{F_p} P$. Hence $M_p/pM_p = \tilde{P}$ and M_p is a $\tilde{C}_{p^{n-1}}$ projective module. □

Let $D = (C \cap D) \oplus D'$ where D' is some complementary summand. Then A

splits into $A = A_{C\cap D} \oplus A_{D'}$. Now

$$\begin{aligned} W^{\perp}/W = H(A) &= A \oplus A^* \\ &= (A_{C\cap D} \oplus A_{D'}) \oplus (A^*_{C\cap D} \oplus A^*_{D'}) \\ &= (A^*_{C\cap D} \oplus A_{D'}) \oplus (A_{C\cap D} \oplus A^*_{D'}) \\ &= H(A^*_{C\cap D} \oplus A_{D'}) \ . \end{aligned}$$

Set $B = A^*_{C\cap D} \oplus A_{D'}$. We claim that $\mathrm{Im}\{H^2(C_{p^n};B) \longrightarrow H^2(C_{p^n};W^{\perp}/W)\} \cap C = \{0\}$. Clearly $\mathrm{Im}\{H^2(C_{p^n};B) \longrightarrow H^2(C_{p^n};W^{\perp}/W)\} = (C\cap D)^* \oplus D'$. If $(a^*,b) \in ((C\cap D)^* \oplus D') \cap C$ then $a^* = 0$. Otherwise there is an $a' \in C \cap D$ such that $a^*(a') \neq 0$. But this contradicts $C \subset C^{\perp}$. Now $b = 0$ because $C \cap D' = \{0\}$.

This finishes the proof of Theorem VI.5.1.

6. Units in $Z[C_{p^k}]$

In the case of a cyclic group of odd prime power order it is possible to give an account of the group $Z(C_{p^k})^*$ by using elementary induction techniques over k.

Let $X \longrightarrow \hat{X}$ be the canonical involution on the integral group ring. We take an action of C_2 on $Z(C_{p^k})^*$ given by $X \longrightarrow \hat{X}^{-1}$. Thus $H^1(C_2;Z(C_{p^k})^*)$ is the group of units for which $X = \hat{X}$ modulo the subgroup of Hermitian squares of units. This group is surely of central importance in the study of Hermitian $Z(C_{p^k})$-valued innerproducts on free $Z(C_{p^k})$-modules. Thus understanding $H^1(C_2;Z(C_{p^k})^*)$ will be our immediate objective.

We denote by T the generator of C_{p^k} so that $\hat{T} = T^{-1}$. As in the last section, let

$$\Delta = I - T^{p^{k-1}}$$

$$\Sigma = I + T^{p^{k-1}} + T^{2p^{k-1}} + \ldots + T^{p^k - p^{k-1}} .$$

Recall that $T^{p^{k-1}}$ generates the cyclic subgroup $C_p \subset C_{p^k}$. We denote the image of

T under the quotient homomorphism $C_{p^k} \to \tilde{C}_{p^{k-1}}$ by τ. There results an epimorphism

$$Z(C_{p^k}) \xrightarrow{E} Z(\tilde{C}_{p^{k-1}}) \to 0$$

the kernel of which is the principal ideal (Δ). Obviously E commutes with canonical involutions. If $\lambda_{p^j} = \exp(2\pi i/p^j)$ then there is also an epimorphism ν: $Z(C_{p^k}) \to Z(\lambda_{p^k}) \to 0$ onto the ring of cyclotomic integers, the kernel of which is (Σ). If complex conjugation is the involution on $Z(\lambda_{p^k})$ then ν also commutes with involutions

(6.1) Lemma: The ideals (Δ), (Σ) are related by

$$(\Delta) \cap (\Sigma) = 0$$

$$(p,\Delta) = (\Sigma,\Delta) .$$

Proof. First of all the two statements

$$(\Delta) = \mathrm{ann}(\Sigma) = \{x \in Z(C_{p^k}) \mid \Sigma x = 0\}$$

$$(\Sigma) = \mathrm{ann}(\Delta) = \{y \in Z(C_{p^k}) \mid \Delta y = 0\}$$

follow from the fact that $Z(C_{p^k})$ is a free C_p-module. Since $\Sigma^2 = p\Sigma$ we find $pI - \Sigma \in \mathrm{Ann}(\Sigma) = \mathrm{im}(\Delta)$. Hence there is $\theta \in Z(C_{p^k})$ (see II.3) with $pI = \Sigma + (pI - \Sigma) = \Delta\theta$. Thus we find $(p,\Delta) = (\Sigma,\Delta)$. For $X \in Z(C_{p^k})$

$$pX = \Sigma X + \Delta\theta X$$

so that if $X \in (\Delta) \cap (\Sigma) = \mathrm{ann}(\Sigma) \cap \mathrm{ann}(\Delta)$ we see $pX = 0$. Hence $(\Delta) \cap (\Sigma) = 0$ as claimed. □

Naturally we have shown

$$0 \longrightarrow Z(C_{p^k}) \xrightarrow{E \oplus \nu} Z(C_{p^{k-1}}) \oplus Z(\Lambda_{p^k})$$

is a monomorphism. We also recognize that the cokernel of $E \oplus \nu$ should be

$$Z(C_{p^k})/(\Delta,\Sigma) = Z(C_{p^k})/(p,\Delta) = F_p(\tilde{C}_{p^{k-1}}) .$$

Take $r: Z(\tilde{C}_{p^{k-1}}) \longrightarrow F_p(\tilde{C}_{p^{k-1}}) \longrightarrow 0$ to be reduction mod p. Also there is a homomorphism $t: Z(\Lambda) \longrightarrow F_p(C_{p^{k-1}})$ which sends λ τ. The kernel is the principal ideal generated by $\nu(\Delta)$.

We arrive at a short exact sequence

$$(6.2) \qquad 0 \longrightarrow Z(C_{p^k}) \xrightarrow{E \oplus \nu} Z(C_{p^{k-1}}) \oplus Z(\Lambda_{p^k}) \xrightarrow{r - t} F_p(C_{p^{k-1}}) \longrightarrow 0 .$$

We are going to use this sequence as the basis for our induction arguments. If $\varepsilon_Z: Z(\cdot) \longrightarrow Z$ is the integral augmentation we shall begin with

(6.3) Lemma: If $0 \leq j < p-1$ <u>then the ideal of elements</u> $X \in Z(C_{p^k})$ <u>for which</u> $\varepsilon_Z(X) \equiv 0 \pmod{j+1}$ <u>is the principal ideal generated by</u> $I + T + \ldots + T^j$.

Proof. Observe $\varepsilon_Z(X) = \varepsilon_Z(E(X))$ so we shall proceed by induction over the exponent of p. Assume $\varepsilon_Z(X) \equiv 0 \pmod{j+1}$, then by induction there is $y \in Z(C_{p^{k-1}})$ with

$$(I + \tau + \ldots + \tau^j)y = E(x) .$$

The element $1 + \lambda + \ldots + \lambda^j$ is a unit in $Z(\Lambda_{p^k})^*$, hence there is a $w \in Z(\Lambda_{p^k})$ with $w(1 + \lambda + \ldots + \lambda^j) = \nu(X)$. Now in $F_p(\tilde{C}_{p^{k-1}})$ we have

$$(I + \tau + \ldots + \tau^j)r(y) = (I + \tau + \ldots + \tau^j)t(w)$$

and in $F_p(\widetilde{C}_{p^{k-1}})$ we can cancel the unit to conclude $r(y) = t(w)$. So there is a $Y \in Z(C_{p^k})$ with $E(Y) = y$, $\nu(Y) = w$. Now $(I + T + \ldots + T^j)Y = X$ as required. □

It is well known that a unit in $Z(\Lambda_{p^k})^*$ can be expressed as the product of a root of unity with a unit in $Z(\Lambda_{p^k} + \lambda^{-1}_{p^k})^*$. Let us prove the analogous result for $Z(C_{p^k})^*$.

(6.4) Lemma: <u>If</u> $X \in Z(C_{p^k})^*$ <u>then</u> X <u>can be expressed as</u> $T^j Y$ <u>with</u> $0 \leq j \leq p^k - 1$ <u>and</u> $Y = \hat{Y} \in Z(C_{p^k})^*$.

Proof. Clearly we are going to assume there is $y = \hat{y} \in Z(C_{p^{k-1}}))^*$ for which $\tau^i y = E(X)$, $0 \leq i \leq p^{k-1} - 1$. Also there is a unit $u = \bar{u} \in Z(\Lambda_{p^k})^*$ with $\lambda^j v = \nu(X)$, $0 \leq j \leq p^k - 1$. Observe in $F_p(\widetilde{C}_{p^{k-1}})$ that $\tau^i(r(y)) = \tau^j t(u)$ and since $r(y), t(u)$ are both fixed under the canonical involution, $\tau^{j-i} = I$ and $r(y) = t(u)$. Thus $j \equiv i \bmod p^{k-1}$ and there is a $Y \in Z(C_{p^k})$ with $E(Y) = y$, $\nu(Y) = u$. Now $X = T^j Y$ as asserted. □

Our next lemma concerns the units in $F_p(C_{p^k})$. This will use the augmentation $\varepsilon_p : F_p(C_{p^k}) \longrightarrow F_p$.

(6.5) Lemma: <u>An element</u> $\alpha \in F_p(C_{p^k})$ <u>is a unit if and only if</u> $\varepsilon_p(\alpha) \in F_p^*$. <u>The group of units is a direct product</u> $F_p^* \times K$ <u>where</u> K <u>is a finite group of odd order</u>.

Proof. Write $\alpha = \sum f_i \tau^i$, $f_i \in F_p$, then $\alpha^{p^k} = (\sum f_i) I = \varepsilon(\alpha) I$. This proves the first assertion. The epimorphism $F_p(C_{p^k})^* \xrightarrow{\varepsilon_p} F_p^*$ is split by $f \longrightarrow f \cdot I$. Thus K is the subgroup of units with augmentation 1 and K has order $p^{p^k - 1}$. □

Now C_2 acts on $F_p(C_{p^k})^*$ by $X \longrightarrow \hat{X}^{-1}$, and K is a C_2-submodule. The action of C_2 on F_p^* is $f \longrightarrow f^{-1}$. From (VI.6.5) it now follows

(6.6) Lemma: <u>The augmentation induces an isomorphism</u>

$$H^*(C_2;F_p(C_{p^k})^*) \simeq H^*(C_2;F_p^*) .$$

Furthermore

$$H^1(C_2;F_p^*) \simeq F_p^*/F_p^{**}$$

$$H^2(C_2;F_p^*) \simeq \{-1\} \in F_p^* . \quad \square$$

The last indicates the cyclic subgroup of order 2 generated by -1. For future reference we state

(6.7) Lemma: *If* $H \subset F_p(C_{p^k})^*$ *is a* C_2*-submodule for which* $\varepsilon_p(H) \subset F_p^*$ *contains the Sylow* 2-*subgroup of* F_p^* *then* $Z(C_{p^k})^*/H$ *has odd order and*

$$\varepsilon_p: H^*(C_2;H) \simeq H^*(C_2;F_p^*) . \quad \square$$

Now passing to units we want to consider

(6.8) $$1 \to Z(C_{p^k})^* \xrightarrow{E \times \nu} Z(C_{p^{k-1}})^* \times Z(\Lambda_{p^k})^* \xrightarrow{r/t} F_p(C_{p^{k-1}})^* .$$

By r/t we simply mean $(x,u) \to r(x)t(u^{-1})$. We have no idea as to whether r/t is an epimorphism. Simply denote its image by $H \subset F_p(C_{p^{k-1}})^*$. We can apply (VI.6.6) to H because

(6.9) Lemma: *The composition*

$$\varepsilon_p \circ t: Z(\lambda_{p^k} + \lambda_{p^k}^{-1})^* \to F_p^*$$

is an epimorphism.

Proof. Actually $\varepsilon_p \circ t: Z(\Lambda_{p^k}) \to F_p$ is the residue map

$$Z(\Lambda_{p^k}) \to Z(\Lambda_{p^k})/(1-\lambda_{p^k}) = F_p .$$

This identifies 1 and λ. As noted earlier $1+\lambda+\ldots+\lambda^j$ is a unit for $0 \leq j < p-1$ and its image in F_p^* is $j+1$. Since a unit in $Z(\lambda_{p^k})^*$ is the product of a root of unity with a real unit we are done. □

Notice we have really shown

$$H^*(C_2;Z(\lambda_{p^k})^*) \longrightarrow H^*(C_2;H) \simeq H^*(C_2;F_p^*)$$

is an epimorphism. Hence we find

(6.10) <u>Theorem</u>: <u>There is a short exact sequence</u>

$$1 \rightarrow H^*(C_2;Z(C_{p^k})^*) \rightarrow H^*(C_2;Z(\tilde{C}_{p^{k-1}})^*) \times H^*(C_2;Z(\lambda_{p^k})^*) \rightarrow H^*(C_2;F_p^*) \rightarrow 1 . \square$$

We want to explain

$$H^1(C_2;Z(\tilde{C}_{p^{k-1}})^*) \times H^1(C_2\ Z(\lambda_{p^k})^*) \longrightarrow F_p^*/F_p^{**}$$

in other terms. Suppose $x = \hat{x} \in Z(C_{p^k})^*$ and $u = \bar{u} \in Z(\lambda_{p^k})^*$. Now $\varepsilon_p r(x) = \varepsilon_Z(x) \pmod p$, thus $(x,u) \rightarrow \varepsilon_Z(x)u \in (Z(\lambda_{p^k})/(1-\lambda_{p^k}))^*$ modulo squares is the homomorphism.

(6.11) <u>Corollary</u>: <u>The following are equivalent</u>

a) $(cl(x), cl(u))$ <u>lies in the image of</u> $H^1(C_2;Z(C_{p^k})^*) \longrightarrow H^1(C_2;Z(\tilde{C}_{p^{k-1}})) \times H^1(C_2;Z(\lambda_{p^k})^*)$

b) $(\varepsilon_Z(x)u,\sigma)_{\mathfrak{p}_0} = 1$ <u>where</u> $(\ ,\ )_{\mathfrak{p}_0}$ <u>is the Hilbert symbol at the ramified prime in</u> $Z(\lambda_{p^k}+\lambda_{p^k}^{-1})$ <u>and</u> $\sigma = (\lambda_{p^k})^{\frac{p^k+1}{2}} (\lambda_{p^k})^{\frac{p^k-1}{2}} \in Z(\lambda_{p^k}+\lambda_{p^k}^{-1})$.

c) $\varepsilon_Z(x)v$ <u>is negative at an even number of orderings of the maximal real subfield of</u> $Q(\Lambda_{p^k})$. □

The equivalence of b and c is a consequence of Hilbert symbol reciprocity and the fact that $(x,\sigma)_{\mathfrak{p}} = 1$ if x is a unit and $\mathfrak{p}$ is an unramified prime. Remember $\varepsilon_Z(x) = \pm 1$.

We can easily improve this result into a more usable form. For $1 \leq m \leq k$ there is a homomorphism

$$\mathscr{E}_m : Z(C_{p^k}) \longrightarrow Z(\tilde{C}_{p^{k-m}}) + \sum_{k+1-m}^{k} Z(\Lambda_{p^j}) .$$

We take $\mathscr{E}_1 = E$ and assume $\mathscr{E}_m$ is inductively defined. Then there is

$$Z(\tilde{C}_{p^{k-m}}) \oplus \sum_{k+1-m} Z(\Lambda_{p^j}) \xrightarrow{E \oplus Id} Z(\tilde{C}_{p^{k-m-1}}) \oplus Z(\Lambda_{p^{k-m}}) \oplus \sum_{k+1-m} Z(\Lambda_{p^j}).$$

The composition of $\mathscr{E}_m$ with this homomorphism then yields $\mathscr{E}_{m+1}$. For $m = p^k$ we get the embedding of $Z(C_{p^k})$ onto $\mathcal{O} = Z \oplus \sum_1^k Z(\Lambda_{p^j})$ which is the maximal order in the group algebra $Q(C_{p^k}) = Q \oplus \sum_1^k Q(\Lambda_{p^j})$. This induces

$$1 \longrightarrow H^1(C_2; Z(C_{p^k})^*) \longrightarrow H^1(C_2; \mathcal{O}^*) .$$

Let us determine the cokernel. Consider $(\varepsilon, v_1, \ldots, v_k)$ where $\varepsilon = \pm 1$ and $v_j = \bar{v}_j \in Z(\Lambda_{p^j})^*$. Map this to $(Z^*)^k$ by

$$(\varepsilon, v_1, \ldots, v_k) \longrightarrow ((\varepsilon v_1, \sigma_1)_{\mathfrak{p}_0^1}, (\varepsilon v_2, \sigma_2)_{\mathfrak{p}_0^2}, \ldots, (\varepsilon v_k, \sigma_k)_{\mathfrak{p}_0^k})$$

where $\sigma_j = ((\Lambda_{p^j})^{\frac{p^j+1}{2}} - (\Lambda_{p^j})^{\frac{p^j-1}{2}})$ and $\mathfrak{p}_0^j$ is the ramified prime in $Z(\Lambda_{p^j} + \Lambda_{p^j}^{-1})$.

(6.12) <u>Corollary</u>: <u>There is a short exact sequence</u>

$$1 \longrightarrow H^1(C_2; Z(C_{p^k})^*) \longrightarrow H^1(C_2; \mathcal{O}^*) \longrightarrow (Z^*)^k \longrightarrow 1 . \quad \square$$

For example if $(\varepsilon v_1, \sigma)_{\mathfrak{p}_0} = 1$ we can lift $(\varepsilon, v_1, \ldots, v_k)$ to

$(x_1, v_2, \ldots, v_k) \in Z(C_p)^* \oplus \sum_2^k Z(\Lambda_{p^j})$. But $\varepsilon_Z(x) = \varepsilon$. We retreat inductively to $Z(C_{p^k})^*$.

One more observation to conclude the section.

(6.13) Lemma: $H^2(C_2; Z(C_{p^k})^*) \simeq C_2$.

Proof. Write the unit as $T^j Y = X$ with $Y = \hat{Y} \in Z(C_{p^k})^*$. If $X = \hat{X}^{-1}$ then $T^j Y T^{-j} Y = I$ or $Y^2 = I$. To see $Y = \pm I$ assume by way of induction that $E(Y)^2 = I$ implies $E(Y) = \pm I$. Of course $\nu(Y)^2 = 1$ implies $\nu(Y) = \pm 1$. Finally $rE(Y) = t\nu(Y)$ shows the signs agree so $Y = \pm I$. The reader can see the rest of the proof. □

7. The ideal class group

In this section, by analogy with chapter III, we work with the ideal class group of $Z(C_{p^k})$. We have $Z(C_{p^k}) \subset Q(C_{p^k})$ so we think of $Q(C_{p^k})$ as a $Z(C_{p^k})$-module. A submodule $A \subset Q(C_{p^k})$ is called a fractional ideal if and only if there is a unit $x \in Q(C_{p^k})^*$ for which $xA \subset Z(C_{p^k})$. To A we associate $A^{-1} = \{y \mid y \in Q(C_{p^k}), yA \subset Z(C_{p^k})\}$ and we say A is invertible if and only if $AA^{-1} = Z(C_{p^k})$. An ideal is projective if and only if it is invertible. If A is invertible then A^{-1} is also a fractional ideal and $(A^{-1})^{-1} = A$. The invertible fractional ideals form an abelian group. On this group C_2 acts as $A \to \hat{A}^{-1}$. Two invertible fractional ideals A and A_1 are equivalent if and only if there is a unit $x \in Q(C_{p^k})^*$ for which $xA = A_1$. The group of equivalence classes, $\mathscr{C}(C_{p^k})$, is the ideal class group of $Z(C_{p^k})$. It inherits the C_2 action.

Now Swan [Sw1] shows that if P is a finitely generated projective $Z(C_{p^k})$-module then $P \otimes_{Z(C_{p^k})} Q(C_{p^k})$ is free as a $Q(C_{p^k})$ module. Denote the rank of $P \otimes Q(C_{p^k})$ by $rk(P) \in Z$. Thus invertible fractional ideals are exactly the rank 1 projectives. More generally

$$0 \longrightarrow P \longrightarrow P \otimes_{Z(C_{p^k})} Q(C_{p^k}) = V$$

yields, using the exterior power,

$$0 \longrightarrow \Lambda^{rk(P)} P \longrightarrow Q(C_{p^k})$$

so that $\det(P) = \Lambda^{rk(P)} P$ is an invertible fractional ideal.

If $K_0(Z(C_{p^k}))$ is the projective class group then $\tilde{K}_0(Z(C_{p^k}))$ is the kernel of the rank homomorphism and the determinant yields a natural isomorphism ([Sw 1,2]).

$$\tilde{K}_0(Z(C_{p^k})) \simeq \mathscr{C}(C_{p^k}) .$$

The action of C_2 on $\tilde{K}_0(Z(C_{p^k}))$ is by dualizing; that is

$$P^* = \mathrm{Hom}_{Z(C_{p^k})}(P, Z(C_{p^k}))$$

and $(X\varphi)(\cdot) \equiv \varphi(\hat{X}\cdot) = \hat{X}(\varphi(\cdot))$ is the $Z(C_{p^k})$-module structure.

The exact sequence (VI.6.2) suggests the application of the six term Mayer-Vietoris sequence of algebraic K-theory ([M]).

(7.1) Lemma: The kernel of $K_0(Z(C_{p^k})) \longrightarrow K_0(Z(\tilde{C}_{p^{k-1}})) \oplus K_0(Z(\Lambda_{p^k}))$ is a finite group of odd order.

Proof. Since $F_p(\tilde{C}_{p^{k-1}})$ is a finite ring, $K_1(F_p(\tilde{C}_{p^{k-1}})) = F_p(\tilde{C}_{p^{k-1}})^*$, ([M]). Let $H \subset F_p(\tilde{C}_{p^{k-1}})^*$ be the image of

$$K_1(Z(\tilde{C}_{p^{k-1}})) \oplus K_1(Z(\Lambda_{p^k})) \longrightarrow F_p(\tilde{C}_{p^{k-1}})^* .$$

Since $K_1(Z(\Lambda_{p^k})) = Z(\Lambda_{p^k})^*$ and $F_p^* \subset F_p(\tilde{C}_{p^{k-1}})^*$ is in the image of $Z(\Lambda_{p^k})^*$

$F_p(\tilde{C}_{p^{k-1}})^*$ we use (VI.6.5) and (VI.6.9) to conclude that the quotient $K_1(F_p(\tilde{C}_{p^{k-1}}))/H$ is a finite group of odd order. □

We want to pass this over to the kernel of

$$K_0(Z(C_{p^k})) \longrightarrow K_0(\mathcal{O}) = K_0(Z) \oplus \sum_1^k K_0(Z(\Lambda_{p^j})) .$$

Call this kernel J. Filter J as follows. Let $J_0 = \{0\}$ and $J_m = \ker: K_0(Z(C_{p^k})) \longrightarrow K_0(Z(C_{p^{k-m}})) + \sum_{k+1-m}^{k} K_0(Z(\Lambda_{p^j}))$. Here $\mathcal{E}_m$ induces the homomorphism. Then J_{m+1}/J_m embeds into the kernel of

$$K_0(Z(\tilde{C}_{p^{k-m}})) \oplus \sum_{k+1-m}^{k} K_0(Z(\Lambda_{p^j})) \longrightarrow$$

$$K_0(Z(\tilde{C}_{p^{k-m-1}}) \oplus K_0(Z(\Lambda_{p^{k-m}})) \oplus \sum_{k+1-m}^{k} K_0(Z(\Lambda_{p^j}))$$

and so by lemma VI.7.1 J_{m+1}/J_m is a finite group of odd order.

(7.2) Lemma: The kernel of $K_0(Z(C_{p^k})) \longrightarrow K_0(\mathcal{O}) = K_0(Z) \oplus \sum_1^k K_0(Z(\Lambda_{p^j}))$ is a finite group of odd order. □

A finitely generated projective $\mathcal{O}$-module splits into a direct sum $\sum_0^k P_j$ with P_j a finitely generated $Z(\Lambda_{p^j})$-module. (Agree that $Z = Z(\Lambda_{p^0})$.) Now $P \otimes_{\mathcal{O}} Q(C_{p^k}) = \sum_0^k P_j \otimes_{Z(\Lambda_{p^j})} Q(\Lambda_{p^j})$ is free as $Q(C_{p^k})$ module if and only if $rk(P) = \ldots = rk(P_k)$. Thus the kernel of "multirank"

$$\sum_0^k K_0(Z(\Lambda_{p^j})) \longrightarrow Z^k$$

by $(P_0, P_1, \ldots, P_k) \longrightarrow (rkP_1 - rkP_0, rkP_2 - rkP_0, \ldots,)$ is the group which Swan calls $C(\mathcal{O}) \subset K_0(\mathcal{O})$. Swan then shows that $\tilde{K}_0(Z(C_{p^k})) \longrightarrow C(\mathcal{O})$ is an epimorphism ([Sw2]), and we have noted the kernel is a finite group of odd order.

(7.3) Theorem: There is an isomorphism

$$H^*(C_2;\tilde{K}_0(Z(C_{p^k}))) \simeq H^*(C_2;C(\mathcal{O})) \ . \ \Box$$

In the ideal class formalism $\mathcal{C}(C_{p^k}) \simeq \tilde{K}_0(Z(C_{p^k}))$. It is also true that $C(\mathcal{O}) \simeq \prod_0^k \mathcal{C}(j)$, the ideal class groups of the various $Z(\Lambda_{p^j})$. This follows from ([Sw2]) because $C(\mathcal{O})$ is the group of projectives of "constant rank". Actually $\mathcal{C}(C_{p^k}) \longrightarrow \prod_0^k \mathcal{C}(j)$ is directly described by splitting up

$$A \otimes_{Z(C_{p^k})} \mathcal{O} = A\mathcal{O} \subset Q(C_{p^k}) \ .$$

(7.4) Corollary: There is an isomorphism

$$H^*(C_2;\mathcal{C}(C_{p^k})) \simeq \prod_0^k H^*(C_2;\mathcal{C}(j)) \ . \ \Box$$

It will be recalled from chapter III that there is a monomorphism

$$1 \longrightarrow H^1(C_2;\mathcal{C}(j)) \longrightarrow H^1(C_2;Z(\Lambda_{p^j})^*)$$

whose image is those units which are Hermitian squares in $Q(\Lambda_{p^j})^*$ but not in $Z(\Lambda_{p^j})^*$. We can give a similar argument for $Z(C_{p^k})^*$. If $A\hat{A}^{-1} \sim Z(C_{p^k})$ then for some $x \in Q(C_{p^k})^*$ we have $A\hat{A}^{-1} = xZ(C_{p^k})$ so $\hat{x}Z(C_{p^k}) = x^{-1}Z(C_{p^k})$ and $\hat{x}x = U \in Z(C_{p^k})^*$ for which $U = \hat{U}$. This describes a well defined map $[A] \longrightarrow [U]$ of $H^1(C_2,\mathcal{C}(C_{p^k})) \longrightarrow H^1(C_2;Z(C_{p^k})^*)$. Using (VI.6.12) and (VI.7.4) we find

(7.5) Corollary: There is a commutative diagram

$$\begin{array}{ccc} \prod_0^k H^1(C_2;\mathscr{C}(j)) & \xrightarrow{\sim} & \prod_0^k H^1(C_2;Z(\Lambda_{p^j})^*) \\ \downarrow & & \downarrow \\ H^1(C_2;\mathscr{C}(C_{p^k})) & \xrightarrow{\sim} & H^1(C_2;Z(C_{p^k})^*) \end{array} .$$

Thus again $H^1(C_2;\mathscr{C}(C_{p^k}))$ _represents units in_ $Z(C_{p^k})^*$ _which become Hermitian squares in_ $Q(C_{p^k})^*$. □

The group $H^2(C_2;\mathscr{C}(C_{p^k}))$ is interpreted in terms of rank one projectives which carry a $Z(C_{p^k})$ valued Hermitian innerproduct. Briefly, if $A \sim \hat{A}^{-1}$ then there is $x \in Q(C_{p^k})^*$ with $A\hat{A} = xZ(C_{p^k})$. Then $\hat{x}x^{-1} = U \in Z(C_{p^k})^*$ with $U\hat{U} = I$. Thus $U = \pm T^{2i}$ and we replace x by $y = xT^i$ so $\hat{y} = \pm y$ and $A\hat{A} = yZ(C_{p^k})$. To show $\hat{y} = -y$ is not possible look at $\nu(A)\overline{\nu(A)} = \nu(y)Z(\Lambda_{p^k})$. By III.2.1 $\overline{\nu(y)} = -\nu(y)$ is impossible. So $\hat{y} = y$. Now on A

$$[a,a_1] = a\hat{a}_1/y$$

is a Hermitian innerproduct.

There is a group $\mathcal{D}is$ for discriminant, which should be introduced at this point. We consider pair (y,A) for which $\hat{y} = y \in Q(C_{p^k})^*$ and $A\hat{A} = yZ(C_{p^k})$. Such form a group with $(I, Z(C_{p^k}))$ as identity. To obtain $\mathcal{D}is$ we factor out by the subgroup of all pairs of the form $(x\hat{x}, xB\hat{B}^{-1})$ with $x \in Q(C_{p^k})^*$ and B an invertible fractional ideal.

(7.6) __Lemma__: _There is an exact sequence_

$$1 \to H^1(C_2;\mathscr{C}(C_{p^k})) \to H^1(C_2;Z(C_{p^k})^*) \xrightarrow{\alpha} \mathcal{D}is \xrightarrow{\beta} H^2(C_2;\mathscr{C}(C_{p^k})) \to 1$$

where $\alpha([U]) = (U, Z((p^k))$ _and_ $\beta(y,A) = [A]$.

Proof. Suppose $(y,A) \longrightarrow 1 \in H^2(C_2;\mathscr{C}(C_{p^k}))$ then there is a invertible fractional ideal B and an $x \in Q(C_{p^k})^*$ for $A = xB\hat{B}^{-1}$. But then

$$xB\hat{B}^{-1}\,\hat{x}\hat{B}B^{-1} = x\hat{x}Z(C_{p^k}) = yZ(C_{p^k}) .$$

Hence there is $U = \hat{U} \in Z(C_{p^k})^*$ with $Ux\hat{x} = y$. Then $(U,Z(C_{p^k}))(x\hat{x},xB\hat{B}^{-1}) = (y,A)$.

Next suppose $(U,Z(C_{p^k})) = (1,Z(C_{p^k}))$ in $\mathcal{D}is$ where $U = \hat{U} \in Z(C_{p^k})^*$. Then $(U,Z(C_{p^k})) = (x\hat{x},xB\hat{B}^{-1})$. Now $B\hat{B}^{-1}$ equivalent to $Z(C_{p^k})$ means B defines a class in $H^1(C_2;\mathscr{C}(C_{p^k}))$ whose image in $H^1(C_2;Z(C_{p^k})^*$ is the class of U. □

For each $Q(\Lambda_{p^j})$ we may define a group $\mathcal{D}is(j)$ by analogy and obtain

$$1 \longrightarrow H^1(C_2;\mathscr{C}(j)) \longrightarrow H^1(C_2;Z(\Lambda_{p^j})^*) \longrightarrow \mathcal{D}is(j) \longrightarrow H^2(C_2;\mathscr{C}(j)) \longrightarrow 1 .$$

Indeed there is

$$\begin{array}{ccccccccc}
1 \to & H^1(C_2;\mathscr{C}) & \to & H^1(C_2;Z(C_{p^k})^*) & \to & \mathcal{D}is & \to & H^2(C_2;\mathscr{C}) & \to 1 \\
 & \downarrow & & \downarrow & & \downarrow & & \downarrow & \\
1 \to & \prod_0^k H^1(C_2;\mathscr{C}(j)) & \to & \prod_0^k H^1(C_2;Z(\Lambda_{p^j})^*) & \to & \prod_0^k \mathcal{D}is(j) & \to & \prod_0^k H^2(C_2;\mathscr{C}(j)) & \to 1 \\
 & & & \downarrow & & & & & \\
 & & & (Z^*)^k & & & & & \\
 & & & \downarrow & & & & & \\
 & & & 1 . & & & & &
\end{array}$$

From this we can define

$$\prod_0^k \mathcal{D}is(j) \longrightarrow (Z^*)^k \longrightarrow 1$$

and conclude

$$(7.7) \qquad 1 \longrightarrow \mathcal{D}is \longrightarrow \prod_{0}^{k} \mathcal{D}is(j) \longrightarrow (Z^*)^k \longrightarrow 1$$

is exact.

8. Type II innerproducts

In this section π is a finite abelian group of odd order. Suppose that (π, V, b) is an integral representation of π as a group of isometries with respect to a Z-innerproduct space (V, b) which may be either symmetric or skew-symmetric. We can canonically associate a $Z(\pi)$-Hermitian form on V to (V, b) by introducing

$$h_b(v, w) = \sum_{g \in \pi} b(v, gw) g .$$

We note h_b is actually Hermitian or skew-Hermitian according to the symmetry of b. There is $Ad\colon V \longrightarrow Hom_{Z(\pi)}(V, Z(\pi))$ $Ad_V(v) = h_b(v, w)$. This is an isomorphism of $Z(\pi)$-modules. In fact there is $Hom_{Z(\pi)}(V, Z(\pi)) \simeq Hom_Z(V, Z)$ where $\psi \in Hom_{Z(\pi)}(V, Z(\pi))$ is sent into the unique $\varphi \in Hom_Z(V, Z)$ that satisfies

$$\psi(v) = \sum_{g \in \pi} \varphi(g^{-1}v) g .$$

Thus in fact we receive a commutative diagram

$$\begin{array}{ccc} & \overset{Ad}{\nearrow} \; \simeq & Hom_{Z(\pi)}(V, Z(\pi)) \\ V & & \downarrow \; \simeq \\ & \underset{ad}{\searrow} \; \simeq & Hom_Z(V, Z) \end{array}$$

We should be careful not to refer to (V, h_b) as a $Z(\pi)$ innerproduct space unless V is $Z(\pi)$-projective.

Observe that $b(v, gv) = b(g^{-1}v, v) = \pm b(v, g^{-1}v)$. We use this simple fact to show

(8.1) Lemma: If (π, V, b) is symmetric then $\varepsilon_z h_b(v, v) \equiv b(v, v) \pmod 2$ while if skew-symmetric then $\varepsilon_z h_b[v, v] = 0$.

Proof. Since π has odd order we can choose $S \subset \pi \{e\}$ so that $S \cap S^{-1} = \phi$ and $S \cup S^{-1} = \pi - \{e\}$. Then

$$h_b(v,v) = b(v,v)e + \sum_{g \in S} b(v,gv)(g+g^{-1})$$

or

$$h_b(v,v) = \sum_{g \in S} b(v,gv)(g-g^{-1}) .$$

The lemma is clear. □

Now let us come to the explicit question we have in mind. Suppose $A \subset Z(C_{p^k})$ is an invertible integral fractional ideal and $y = \hat{y} \in Q(C_{p^k})^*$ is a unit for which $A\hat{A} = yZ(C_{p^k})$. Is there a Hermitian innerproduct structure on $Z(C_{p^k}) \oplus A$ for which the associated symmetric integral innerproduct, b, is even (of Type II)?

We think, by analogy with chapter III of a Hermitian matrix

$$\begin{bmatrix} \alpha & \beta \\ \hat{\beta} & y^{-1}\gamma \end{bmatrix}$$

where $\alpha = \hat{\alpha}$, $\gamma = \hat{\gamma}$ lie in $Z(C_{p^k})$ and $\beta \in A^{-1}$. If $y^{-1}\alpha\gamma - \beta\hat{\beta} = y^{-1}U$, with $U = \hat{U} \in Z(C_{p^k})^*$ then a $Z(C_{p^k})$-Hermitian inner-product on $Z(C_{p^k}) \oplus A$ is given by

$$h((X,X_1),(Y,Y_1)) = \alpha X\hat{Y} + \hat{\beta} X \hat{Y}_1 + \beta X_1 \hat{Y} + y^{-1}\gamma X_1 \hat{Y}_1 .$$

According to (VI.8.1) we also want

$$\varepsilon(\alpha)\varepsilon(X^2) + 2\varepsilon(X)\varepsilon(\beta X_1) + \varepsilon(y^{-1})\varepsilon(\gamma)\varepsilon(X_1^2) \equiv 0 \pmod 2$$

for all (X,X_1) in $Z(C_{p^k}) \oplus A$. Surely we need $\varepsilon(\alpha) \equiv 0 \pmod 2$. Let us look at $\varepsilon(\gamma)$. First $A \subset Z(C_{p^k})$ so that $\varepsilon(A) \subset Z$ is an ideal which is generated by

$N > 0$. Now $yZ(C_{p^k}) = A\hat{A}$ so $y = \sum a_i b_i$ with $\hat{b}_i \in A$, and hence $\varepsilon(y)$ is divisible by N^2. On the other hand there is $a_0 \in A$ with $\varepsilon(a_0) = N$ and $a_0\hat{a}_0 = yX$ with $X = X \in Z(C_{p^k})$. Thus $N^2 = \varepsilon(y)\varepsilon(X)$, and we conclude $\varepsilon(y) = \pm N^2$. Now if $\varepsilon(y^{-1})\varepsilon(\gamma)\varepsilon(a_0)^2$ is to be even we see that $\varepsilon(\gamma) \equiv 0 \pmod 2$ also, and $\varepsilon(\gamma) \equiv 0 \pmod 2$, $\varepsilon(\alpha) \equiv 0 \pmod 2$ is the necessary and sufficient condition.

Return to $y^{-1}\alpha\gamma - \beta\hat{\beta} = y^{-1}U$. Multiply by y to obtain $\alpha\gamma - y\beta\hat{\beta} = U$. Now on A^{-1} we see $\langle\beta,\beta_1\rangle = y\beta\hat{\beta}_1$ is a Hermitian innerproduct. We claim there is $\beta \in A^{-1}$ for which $y\beta\hat{\beta}$ has <u>odd</u> augmentation. If this were not true then the Z-innerproduct underlying $\langle\beta,\beta_1\rangle$ would be a Type II on A which has rank p^k, odd, over Z.

Given $\beta \in A^{-1}$ with $\varepsilon(y\beta\hat{\beta}) \in Z$ odd we select any unit for which $\varepsilon(U) + \varepsilon(y\beta\hat{\beta}) \equiv 0 \pmod 4$. Refer to (VI.6.2) to see that

$$U + y\beta\hat{\beta} \in (T + T^{-1})^2 Z(C_{p^k}) .$$

If $U + y\beta\hat{\beta} = (T + T^{-1})^2 t$, $\hat{t} = t \in Z(C_{p^k})$ then

$$\begin{bmatrix} (T+T^{-1})t & \beta \\ \hat{\beta} & y^{-1}(T+T^{-1}) \end{bmatrix}$$

is exactly what we want

(8.2) <u>Lemma</u>: <u>Let</u> $U = \hat{U} \in Z(C_{p^k})^*$ <u>then there is a</u> $\begin{bmatrix} \alpha & \beta \\ \beta & y^{-1}\gamma \end{bmatrix}$ <u>with determinant</u> $y^{-1}U$ <u>and</u> $\varepsilon(\alpha) \equiv \varepsilon(\gamma) \equiv 0 \pmod 2$ <u>if and only if</u> $-\varepsilon(yU) = N^{-2}$.

Proof. As we noted there is $\beta \in A^{-1}$ with $\varepsilon(y\beta\hat{\beta}) = \varepsilon(y)\varepsilon(\beta)^2$ an odd integer. This integer is ± 1 mod 4 according to $\varepsilon(y) = \pm N^2$. Now U must satisfy $\varepsilon(U) + \varepsilon(y)\varepsilon(\beta^2) \equiv 0$ mod 4 so $\varepsilon(U) = \mp 1$ according to $\varepsilon(y) = \pm N^2$. The lemma follows. □

Actually we can extend this idea as follows. If C_2 acts on $Q(C_{p^k})^*$ by

$x \longrightarrow \hat{x}^{-1}$ then $H^1(C_2;Q(C_{p^k})^*) \simeq \prod_0^k H^1(C_2;Q(\Lambda_{p^j})^*)$ receives the discriminant invariant. That is, suppose V is a free $Q(C_{p^k})$-module of rank $2m$ then to any Hermitian innerproduct on V there is associated a discriminant in $H^1(C_2;Q(C_{p^k})^*)$, that is, $(-1)^m$ determinant. There is a well defined augmentation $\varepsilon\colon H^1(C_2;Q(C_{p^k})^*) \longrightarrow Q^*/Q^{**}$.

Now suppose (F,h) is a free $(Z(C_{p^k})$-module with a $Z(C_{p^k})$-valued Hermitian innerproduct that satisfies $\varepsilon(h(v,v)) \equiv 0 \pmod 2$. Then the rank of F over $Z(C_{p^k})$ is $2m$. With respect to a basis represent h by a Hermitian matrix $[M_{i,j}]$. This has $\det[M_{i,j}] = U = \hat{U} \in Z(C_{p^k})^*$. Now $[\varepsilon(M_{i,i})]$ is a symmetric integral matrix with determinant $\varepsilon(U) = \pm 1$ and $\varepsilon(M_{i,j}) \equiv 0 \pmod 2$. It is known that the determinant of $[\varepsilon(M_{i,j})]$ is $+1$. Thus $\varepsilon(\mathrm{dis}(F,h)) = 1 \in Q^*/Q^{**}$.

We can extend to the case where (P,h) is only projective over $Z(C_{p^k})$. First $P \otimes_{Z(C_{p^k})} Q(C_{p^k})$ is free and has a discriminant in $H^1(C_2;Q(C_{p^k})^*)$. In fact if $2m$ is the rank then the induced form on $\Lambda^{2m}(P \otimes_{Z(C_{p^k})} Q(C_{p^k})) \simeq Q(C_{p^k})$ is multiplication by an element $y = \hat{y} \in Q(C_{p^k})^*$ and the discriminant is $(-1)^m y \in H^1(C_2;Q(C_{p^k})^*)$.

But $P = F \oplus A$, according to Swan, with F free and $A \subset Z(C_{p^k})$ an invertible integral ideal ([Swl]). Thus $\Lambda^{2m}P = \Lambda^{2m-1}F \otimes A$. So on A, $\langle a, a_1\rangle = y\, a\, \hat{a}_1$ is a Hermitian $Z(C_{p^k})$ form. Hence $y A \hat{A} = Z(C_{p^k})$.

Now $P \oplus Z(C_{p^k}) \oplus \hat{A}$ is free. There is on $Z(C_{p^k}) \oplus \hat{A}$ a Hermitian innerproduct h' with $\varepsilon_Z h'(v,v) = 0 \pmod 2$. According to (VI.8.2) $\mathrm{dis}(Z(C_{p^k}) \oplus A)$ has augmentation a square in Q^{**}. By our earlier remark so does $\mathrm{dis}(P \oplus Z(C_{p^k}) \oplus \hat{A}, h \oplus h')$. Hence

(8.3) Lemma: *If (P,h) is a finitely generated projective $Z(C_{p^k})$-module together with a Hermitian $Z(C_{p^k})$ innerproduct that satisfies $h(v,v) \equiv 0 \pmod 2$ then $\varepsilon(\mathrm{dis}(P,h)) = 1 \in Q^*/Q^{**}$.* □

9. Relation to surgery obstruction groups

For the cyclic group of odd prime power the results of this chapter can be applied to the surgery obstruction groups $L_0^P(C_{p^k})$ and $L_0^h(C_{p^k})$. In this way we recover previously announced results about these groups. With this we shall close out these notes.

We comment that if P is a projective $Z(C_{p^k})$-module with a Z-valued innerproduct structure of type II with respect to which C_{p^k} acts as a group of isometries then the associated $Z[C_{p^k}]$-valued Hermitian innerproduct on P is

$$[x,y] = \sum_{j=0}^{p^k-1} (x,T^j y)T^j .$$

The intersection form is

$$\mu(x) = \frac{(x,x)}{2} + \sum_{j=1}^{p^k-1/2} (x,T^j x)T^j$$

and we think of μ taking values in $Z[C_{p^k}]/(X-\hat{X}|X \in Z[C_{p^k}])$. The Hermitian and the Z-innerproducts uniquely determine each other and the intersection form μ as well.

Thus we see there is a natural homomorphism $L_0^P(C_{p^k}) \longrightarrow W_0(Z,C_{p^k})$.

(9.1) Lemma: The homomorphism

$$L_0^P(C_{p^k}) \longrightarrow W_0(Z,C_{p^k})$$

is a monomorphism and $L_0^P(C_{p^k})$ has no torsion.

Proof. Immediate from VI.5.1. $\square$

There is a homomorphism

$$L_0^P(C_{p^k}) \longrightarrow \mathfrak{D}is .$$

It works like this. If $P \in L_0^p(C_{p^k})$ then $P \otimes_{Z[C_{p^k}]} Q[C_{p^k}]$ is free of rank $2m$. Then $\Lambda^{2m} P = A \subset Q[C_{p^k}]$ is an invertible fractional ideal and the Hermitian innerproduct on A induced by that on P can be identified with $[a,a'] = a\hat{a}'/y$ for a unique unit $y = y^{-1} \in Q[C_{p^k}]^*$ such that $yZ[C_{p^k}] = A\hat{A}$. Then $\mathrm{dis}(P) = ((-1)^m y, A) \in \mathcal{D}is$.

The group $L_0^h(C_{p^k})$, the obstruction group to surgery up to homotopy equivalence, is based on Hermitian innerproduct space structures on stably free $Z[C_{p^k}]$-modules. Hence there is a forgetful homomorphism $L_0^h(C_{p^k}) \quad L_0^p(C_{p^k})$. In addition there is a discriminant homomorphism $L_0^h(C_{p^k}) \quad H^1(C_2;Z[C_{p^k}]^*)$. Represent an element of $L_0^h(C_{p^k})$ on a free $Z[C_{p^k}]$-module. The determinant of the resulting Hermitian matrix is a unit $U = \hat{U}$ and $\mathrm{cl}((-1)^m U) \in H^1(C_2;Z[C_{p^k}]^*)$ is the discriminant. Referring to VI.7.6 we can show

(9.2) <u>Theorem</u>: <u>There is a commutative diagram</u>

$$\begin{array}{ccccccccc}
0 \to & H^1(C_2;\tilde{K}_0(Z[C_{p^k}])) & \to & L_0^h(C_{p^k}) & \to & L_0^p(C_{p^k}) & \to & H^2(C_2;\tilde{K}_0(Z[C_{p^k}])) & \to 0 \\
 & \downarrow \cong & & \downarrow & & \downarrow & & \downarrow \cong & \\
1 \to & H^1(C_2;\mathcal{C}(C_{p^k})) & \to & H^1(C_2;Z[C_{p^k}]^*) & \to & \mathcal{D}is & \to & H^2(C_2;\mathcal{C}(C_{p^k})) & \to 0 \ .
\end{array}$$

Proof. The only point requiring guidance is that $L_0^p(C_{p^k}) \to H^2(C_2;\tilde{K}_0(Z[C_{p^k}]))$ is an epimorphism. This is exactly the construction in section 8 of this chapter. We comment also that if V is a projective $Z[C_{p^k}]$ then V represents a class in $H^1(C_2;\tilde{K}_0(Z[C_{p^k}]))$ if and only if $V \oplus V^*$ is stably free. Then the hyperbolic form on $V \oplus V^*$ defines the element in $L_0^h(C_{p^k})$. The rest is left to the reader and is elementary. $\square$

This is a special case of a result announced by Bak [B]. We see clearly just how $L_0^h(C_{p^k})$ may have torsion. By contrast Wall has shown $L_0^s(C_{p^k})$ has no torsion.

The assumption that the adjoint is a simple isomorphism of stably based modules makes the difference.

References

[A] Alexander, J.P., A $W(Z_2)$ invariant for orientation preserving involutions, Proc. Amer. Math. Soc., 51(1975), 455-460.

[ACHV] Alexander, J.P., P.E. Conner, G.C. Hamrick and J.W. Vick, Witt classes of integral representations of an abelian p-group, Bull. Amer. Math. Soc., 80(1974), 1179-1182.

[AHV1] Alexander, J.P., G.C. Hamrick, and J.W. Vick, Bilinear forms and cyclic group actions, Bull. Amer. Math. Soc., 80(1974), 730-734.

[AHV2] Alexander, J.P., G.C. Hamrick and J.W. Vick, Linking forms and maps of odd prime order, Trans. Amer. Math. Soc., 221(1976), 169-185.

[AS] Atiyah, M.F. and G.B. Segal, The index of elliptic operators, II, Ann. of Math., 87(1968), 531-545.

[B] Bak, A., The computation of surgery groups of odd torsion groups, Bull. Amer. Math. Soc., 80(1974), 1113-1116.

[B-S] Bak, A. and W. Scharlau, Grothendieck and Witt groups of orders and finite groups, Inventiones Math., 23(1974), 207-240.

[C-E] Cartan, H. and S. Eilenberg, Homological Algebra, Princeton University Press, Princeton, New Jersey, (1956).

[Ca-F] Cassels, J.W.S., A. Frohlich, Algebraic Number Theory, Thompson Book Company, Washington, D.C., 1967, p. 113.

[C1] Conner, P.E., Seminar on Periodic Maps, Lecture Notes in Math. 46, Springer-Verlag, Heidelberg, (1967).

[C2] Conner, P.E., Corollaries to the AHV-theorem, Semigroup Forum, 9(1974/75), 310-317.

[C-F] Conner, P.E. and E.E. Floyd, Maps of odd period, Annals of Math., (2), 84(1966), 132-156.

[C-R] Conner, P.E. and F. Raymond, A quadratic form on the quotient of a periodic map, Semigroup Forum, 7(1974), 310-333.

[Cu-R] Curtis, C.W. and I. Reiner, Representation Theory of Finite Groups and Associative Algebras, Interscience Publishers, New York (1962).

[D1] Dress, A., Induction and structure theorems for Grothendieck and Witt rings of orthogonal representations of finite groups, Bull. Amer. Math. Soc., 79(1973), 741-745.

[D2] Dress, A., A note on Witt rings, Bull. Amer. Math. Soc., 19(1972), 738-740.

[D3] Dress, A., Induction and structure theorems for orthogonal representations of finite groups, Annals of Math., (2), 102(1975), 291-325.

[Dy] Dyer, E., Cohomology Theories, W.A. Benjamin, New York, (1969).

[F] Frolich, A., Orthogonal and symplectic representations of groups, Proc. London Math., (3), 24(1972), 470-506.

[G] Gibbs, D.E., Some results on orientation preserving involutions, Trans. Amer. Math. Soc., 218(1976), 321-332.

[L] Landherr, W., Ä-quivalenz Hermitescher Formen uber einen beliebigen algebraischen Zahlkörper, Abh. Math. Sem., Hamburg Univ., 11(1935), 245.

[LJM] Lopez de Medrano, S., Cobordism of diffeomorphisms of (k-1)-connected 2k manifolds, Lecture Notes in Mathematics 298, Springer-Verlag, Heidelberg, (1974).

[M] Milnor, J.W., Introduction to Algebraic K-Theory, Ann. of Math. Studies 72, Princeton Univ. Press, Princeton, N.J. (1971).

[M-H] Milnor, J.W. and D. Husemoeller, Symmetric Bilinear Forms, Ergebnisse series, Springer-Verlag, Heidelberg (1974).

[P] Pardon, W., Local surgery and applications to the theory quadratic forms, Bull. Amer. Math. Soc., 82(1976), 131.

[S] Serre, J.P., Homologie singuliere des espaces fibres, Ann. of Math., (2), 54(1951), 425-505.

[Sf] Stoltzfus, N.W., Unravelling the integral knot concordance group, Mem. Amer. Math. Soc., (to appear).

[St] Stong, R.E., Notes on Cobordism Theory, Princeton University Press, Princeton, N.J. (1968).

[Sw1] Swan, R.G., Induced representations and projective modules, Ann. of Math., (2), 71(1960), 552-557.

[Sw2] Swan, R.G., The Grothendieck ring of a finite group, Topology 2(1963), 85-110.

[Sw3] Swan, R.G., K-Theory of Finite Groups and Orders, Lecture Notes in Mathematics 149, Springer-Verlag, Heidelberg (1970).

[Wa1] Wall, C.T.C., On the axiomatic foundations of the theory of Hermitian forms, Proc. Camb. Phil. Soc., 67(1970), 243-250.

[Wa2] Wall, C.T.C., On the classification of Hermitian forms I. Rings of algebraic integers, Comp. Math., 22(1970), 425-451.

[Wa3] Wall, C.T.C., On the classification of Hermitian forms II, III, IV, V, Inventiones Math., 18(1972), 119-141, 19(1973), 59-71, 23(1974), 241-260; 261-288.

[Wa4] Wall, C.T.C., Classification of Hermitian forms VI. Group rings, Ann. of Math., (2), 103(1975), 1.

[We] Weiss, E., Algebraic Number Theory, McGraw-Hill, New York (1963).

[W] Witt, E., Theorie der quadratischen Formen in beliebigen Korpern, J. Reine u. Angew. Math., 176(1937), 31-44.

In our discussion of Grothendieck groups in chapter 4 we include $GR(Q/Z,\pi)$. A comment about the duality involution on this group is in order. If (π,A) is a representation of π on a finite abelian group it would be natural to regard the dual as $(\pi,\mathrm{Hom}_Z(A,Q/Z))$ with π acting by $(g\varphi)(a) \equiv \varphi(g^{-1}a)$. This does define an involution on $GR(Q/Z,\pi)$ which is compatible with the isomorphism $GR(Q/Z,\pi) \simeq \sum GR(F_p,\pi)$. However there is also a natural homomorphism $GR(Q/Z,\pi) \to GR(Z,\pi)$. This can only be made compatible with duality on $GR(Z,\pi)$ if we use $(\pi,A) \to -(\pi,\mathrm{Hom}_Z(A,Q/Z))$ as the duality operator on $GR(Q/Z,\pi)$. In fact, in section 1 of chapter IV we employ this second convention. However as soon as we begin to relate $GR(Q/Z,\pi)$ to $\sum GR(F_p,\pi)$ we switch back to the first, and more natural, definition of duality as $(\pi,A) \to (\pi,\mathrm{Hom}_Z(A,Q/Z))$.

Symbols

1.	Q	rational number field
2.	Z	rational integers
3.	F_p	field of rational integers mod p
4.	Z_m	ring of rational integers mod m
5.	$Z_{(p)}$	rational integers localized at p
6.	D	Dedekind domain or a field
7.	S	algebra over D with involution
8.	π	finite group
9.	$W_*(D,\pi)$	Witt ring of actions of π as a group of isometries with respect to a D-module innerproduct
10.	$D[t]$	polynomials over D
11.	$D[t,t^{-1}]$	finite Laurant series over D
12.	$GR(D,\pi)$	Grothendieck representation ring of π over D
13.	$\text{Burn}(\pi)$	Burnside ring of π
14.	$(\pi,V) \longrightarrow (\pi,\text{Hom}_D(V,Z)) - \text{Ext}_D^1(D,V)$	duality involution on $GR(D,\pi)$
15.	$H^*(C_2;GR(D,\pi))$	cohomology with respect to duality involution
16.	C_n	cyclic group of order n
17.	λ_n	$\exp(2\pi i/n)$, primitive n^{th} root of unity
18.	$Q(\lambda_n)$	n^{th} cyclotomic number field
19.	$\Phi_n(t)$	n^{th} cyclotomic polynomial
20.	$Z(\lambda_n)$	algebraic integers in $Q(\lambda_n)$
21.	$Q(\lambda_n+\lambda_n^{-1})$	maximal real subfield in $Q(\lambda_n)$

22.	$Z(\lambda_n+\lambda_n^{-1})$	Integers in $Q(\lambda_n+\lambda_n^{-1})$
23.	$\mathcal{P}$	prime ideal in $Z(\lambda_n)$
24.	$\mathfrak{p}$	prime ideal in $Z(\lambda_n+\lambda_n^{-1})$
25.	$\mathcal{O}(\mathcal{P})$	local ring of integers
26.	$m(\mathcal{P})$	maximal ideal in $\mathcal{O}(\mathcal{P})$
27.	Π	local uniformizer (generator) of $m(\mathcal{P})$
28.	$(y,\sigma)_{\mathfrak{p}}$	Hilbert symbol
29.	$(\frac{p}{q})$	Legendre symbol
30.	$\mathcal{C}$	ideal class group (of $Q(\lambda_n)$)
31.	$\varphi(\cdot)$	Euler φ-function
32.	$G(E\|F)$	Galois group of $E\|F$
33.	$G(p,Q(\lambda_n)/Q)$	Artin symbol at the rational prime p in the cyclotomic extension
34.	$Q(\lambda_n)/Z(\lambda_n)$	quotient regarded as $Z(\lambda_n)$-module
35.	Q/Z	quotient as a Z-module
36.	$W_*(Q/Z,\pi)$	Witt group of finite forms with π-operating
37.	$R_{\pi}^{\pi'}$, $I_{\pi'}^{\pi}$	restriction and induction homomorphisms
38.	$E_{\pi}\subset H^2(C_2;GR(D,\pi))$	image of $\sum_{\pi'\subset\pi} H^2(C_2;GR(D,\pi'))\xrightarrow{\oplus I_{\pi'}^{\pi}} H^2(C_2;GR(D,\pi))$ summing over all proper subgroups
39.	$\mathcal{H}_o(Z(\lambda_n))$, $\mathcal{H}_2(Z(\lambda_n))$	Witt groups of Hermitian and skew Hermitian forms over $Z(\lambda_n)$